NATIONAL RESEARCH COUNCIL OF CANADA
MONOGRAPH PUBLISHING PROGRAM

Sea Buckthorn

(*Hippophae rhamnoides* L.): Production and Utilization

Thomas S.C. Li and Thomas H.J. Beveridge
Pacific Agri-Food Research Centre
Agriculture and Agri-Food Canada
Summerland, BC , Canada V0H 1Z0

with contributions by
B.D. Oomah
Agriculture and Agri-Food Canada
Food Research
Summerland, BC V0H 1Z0

W.R. Schroeder
Agriculture and Agri-Food Canada
The Prairie Farm Rehabilitation Administration
Shelterbelt Centre Indian Head, SK S0G 2K2

and E. Small
Agriculture and Agri-Food Canada
Biodiversity (Mycology and Botany) - ECORC (Ottawa)
Ottawa, ON K1A 0C6

NRC Research Press
Ottawa 2003

ISBN 0-660-19007-9
NRC No. 46317

National Library of Canada cataloguing in publication data

Li, Thomas S.C.

Sea buckthorn (*Hippophae rhamnoides* L.) : production and utilization

Includes an abstract in French.
Includes bibliographical references.
Issued by National Research Council Canada.
ISBN 0-660-19007-9

1. Sea buckthorn – Canada.
2. *Hippophae rhamnoides* – Canada.
I. Beveridge, Thomas H.J.
II. National Research Council Canada.
III. Title.

SB386.S4L5 2003   634.9'7386   C2003-980079-2

Correct citation for this publication: Li, T.S.C., and Beveridge, T.H.J. (with contributions by B.D. Oomah, W.R. Schroeder, and E. Small). 2003. Sea Buckthorn (*Hippophae rhamnoides* L.): Production and Utilization. NRC Research Press, Ottawa, Ontario. 133p.

Table of Contents

Foreword

Numerous plant species have been used by people, including more than 20 000 for medicine and as many for food. Despite this impressive choice, only a few hundred species are important enough to be traded internationally, and almost all of these have saturated the market to the point that additional investment in them is risky. The agricultural sector has been desperately seeking and exploring new crops for the past half century, but the need seems greater than ever. The appearance of a "Cinderella crop" with great potential is therefore a most welcome development. Indeed, although there probably is no such thing, one is tempted to call sea buckthorn a "miracle plant" in view of its many attractive features.

Sea buckthorn is like an extremely talented entertainer who has worked in obscurity for many years and finally becomes an "overnight success." In North America, most people know it simply as an attractive ornamental shrub with silvery deciduous leaves and bright orange berries. In fact, this "sleeping giant" has been used for more than a thousand years in China. In modern China, fruit is harvested from over 1 million ha of wild sea buckthorn and almost 300 000 ha of cultivated plants, and many sea buckthorn products are made. In the last half century in the Old World, sea buckthorn has been turned into a significant fruit crop, but only in the last decade have North Americans seriously initiated commercial cultivation.

Sea buckthorn has the potential to meet three key marketplace demands of modern times. The first is that food should be nutritional, not just tasty. Sea buckthorn fruits are among the most nutritious and vitamin-rich of all berries, with exceptionally high content of antioxidants, now appreciated to be key to countering the effects of aging and disease. While the berries are too acidic to eat fresh for most palates, the flavor has been likened to a combination of passionfruit and pineapple, and the fruit makes superb mixed juices, jellies, marmalades, sauces, and liqueurs. It not only tastes good, if it is good for you.

Second, society has become acutely conscious of the need to promote health, and consumer goods that do this tend to be highly successful. Not only does sea buckthorn have health-giving nutritional value, it also has healthful cosmetic properties because the berries are high in essential fatty acids. These are important for the maintenance of a healthy complexion, and there is great potential for products that prevent damage to the skin from wind and sun.

Third, an important current issue is the welfare of the environment. Sea buckthorn has acquired an enormous reputation for preventing soil erosion, and has been widely planted for soil conservation and reclamation, notably in China and Canada. This is one of the uncommon plants that has a partnership in its roots with bacteria that provide nitrogen, the key element for plant growth. Therefore, unlike most crops, relatively little fertilizer is required, and there is lessened danger of polluting the environment with agricultural run-off. All things considered, sea buckthorn is one of the most ecologically desirable of agricultural plants.

The present book meets the urgent need for a practical, comprehensive agricultural guide for growing and harvesting sea buckthorn, as well as selecting superior varieties and developing the industry. This monograph is a gold mine of essential and fascinating information, and will undoubtedly be the standard reference on the subject for many years. The authors, internationally renowned scientists with extensive research experience on sea buckthorn, are to be congratulated for this outstanding and timely contribution.

Ernest Small, Ph.D.
Principal Research Scientist
Agriculture and Agri-Food Canada
Eastern Cereal and Oilseed Research Centre
Ottawa, Canada

Preface

Sea buckthorn is hardly a household name in North America but it has provided medicines and sustenance to people in Europe and Asia, especially China for hundreds if not thousands of years. This plant was introduced to Canada about 50 years ago and has been used as a shelterbelt species on the Canadian prairies and for land reclamation and restoration because of its nitrogen fixing abilities. With the recognition of the health benefits which can accrue from the consumption of a diet rich in nutrients such as phytosterols for heart disease prevention and antioxidants such as vitamin C, tocopherols, carotenes, and flavonoids, sea buckthorn has become of considerable interest as a new and valuable crop. It is uniquely rich in all of these substances making it an all-in-one phyto-medicinal cabinet.

The possibility that sea buckthorn could provide the base of a new industry for Canada requires that the best cultural practices for growing the plant be known and understood. Furthermore, the fruit from the plant, while serving a small fresh market, is processed into an array of products, including juices, vegetable oils, vitamin supplements, teas, fruit leathers, and many other useful and nutritious products. This book has been prepared with this intention in mind, that is to provide a starting knowledge base for the creation of an industry for Canada. It provides in a single source the knowledge necessary to grow, harvest, store, and process the fruit and leaves into an array of products. As well, the composition of most of these products is documented and the relationship between composition and nutritional value explained. Specific health benefits are documented when they are known.

Looking to a bright future for a nascent industry!

Thomas S.C. Li
Thomas H.J. Beveridge

Chapter 1. Introduction

Thomas S. C. Li

Agriculture and Agri-Food Canada, Pacific Agri-Food Research Centre
Summerland, British Columbia, Canada V0H 1Z0

Sea buckthorn (*Hippophae rhamnoides* L.) is a hardy, deciduous shrub belonging to the family Elaeagnaceae (Rousi 1971). It bears yellow or orange fruits (Fig. 1.1) that have been used for centuries in both Europe and Asia for food, therapeutic, and pharmaceutical purposes (Bailey and Bailey 1978). Sea buckthorn occurs as a native plant distributed widely throughout temperate zones between 27 and 69EN latitude and 7EW and 122EE longitude (Rousi 1971; Pan et al. 1989) including China, Mongolia, Russia, United Kingdom, France, Denmark, Netherlands, Germany, Poland, Finland, Sweden, and Norway (Fig. 1.2) (Pearson and Rogers 1962; Wahlberg and Jeppsson 1990; Yao and Tigerstedt 1995; Li and Schroeder 1996). In ancient Greece, leaves and young branches were added to horse fodder which resulted in weight gain and healthy shine to the horse's coat. The generic Latin name, *Hippophae*, derives from this practice as it means shining horse (Lu 1992).

The medicinal value of sea buckthorn was first recorded in the Tibetan medical classic "rGyud Bzi" in the 8th century (Xu 1994). Some early scientific research in Russia reported sea buckthorn as the source of medicinal and nutritional products (Abartene and Malakhovskis 1975). A sea buckthorn industry has been thriving in Russia since the 1940's when scientists there began investigating the biologically active substances found in the fruit, leaves, and bark (Beldean and Leahu 1985). The first Russian factory producing products from sea buckthorn was located in Bisk. These products contributed to the diet of Russian cosmonauts and were used as a cream for protection from cosmic radiation (Eliseev 1976; Centenaro et al. 1977). The Chinese experience with sea buckthorn products is more recent, although traditional uses of this crop date back many centuries. Research and plantation establishment in China were initiated in the 1980's. Since 1982 over 500 000 ha of sea buckthorn have been planted and 150 processing factories have been established, producing over 200 products (Li, personal observation). During the last decade, sea buckthorn has attracted considerable attention from researchers around the world, most recently in North America, mainly for its nutritional and medicinal value.

Sea buckthorn can be cultivated, but fails to set fruit at an altitude of 3900 m (Rousi 1965; Rousi 1971). In Russia, large, native populations grow at altitudes of 1200–2000 m above sea level (Eliseev and Fefelov 1977). It can withstand temperatures from –43°C to +40°C (Lu 1992). Sea buckthorn is considered to be drought resistant (Heinze and Fiedler 1981; Kondrashov and Sokolova 1990); however, most natural populations grow in areas receiving 400–600 mm of annual precipitation. Myakushko et al. (1986) recommended

Fig. 1.1. Sea buckthorn plant.

Fig. 1.2. The distribution of sea buckthorn in Europe and Asia. From Li and Schroeder 1996. (With permission).

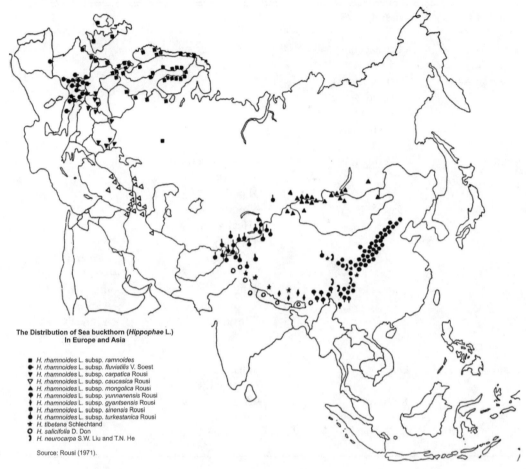

The Distribution of Sea buckthorn (*Hippophae* L.)
In Europe and Asia

■ *H. rhamnoides* L. subsp. *ramnoides*
◆ *H. rhamnoides* L. subsp. *fluviatilis* V. Soest
▼ *H. rhamnoides* L. subsp. *carpatica* Rousi
▽ *H. rhamnoides* L. subsp. *caucasica* Rousi
▲ *H. rhamnoides* L. subsp. *mongolica* Rousi
♥ *H. rhamnoides* L. subsp. *yunnanensis* Rousi
♦ *H. rhamnoides* L. subsp. *gyantsensis* Rousi
● *H. rhamnoides* L. subsp. *sinensis* Rousi
♠ *H. rhamnoides* L. subsp. *turkestanica* Rousi
★ *H. tibetana* Schlechtand
○ *H. salicifolia* D. Don
♪ *H. neurocarpa* S.W. Liu and T.N. He

Source: Rousi (1971).

that sea buckthorn not be grown on dry soils, and Lu (1992) noted the need for irrigation in regions receiving less than 400 mm of precipitation per year. Some species or sub-species of sea buckthorn can endure inundation, but cannot be grown on heavy, water-logged soils (Myakushko et al. 1986), although they take up water rapidly (Heinze and Fiedler 1981). Sea buckthorn rapidly develops an extensive root system, and is therefore an ideal plant for preventing soil erosion (Cireasa 1986; Yao and Tigerstedt 1994), and has been used in land reclamation in Russia and China (Kluczynski 1979; Egyed-Balint and Terpo 1983; Schroeder and Yao 1995). Its ability to fix nitrogen and conserve other essential nutrients make it particularly valuable (Andreeva et al. 1982; Akkermans et al. 1983; Dobritsa and Novik 1992).

Sea buckthorn can be used for many purposes (Table 1.1) and, thus, has considerable economic potential. A natural sea buckthorn habitat can yield 0.75–1.5 t/ha of fruits (Lu 1992). Wolf and Wegert (1993) reported a yield of 5 t/ha from an orchard plantation in

Table 1.1. Possible uses for components in different sea buckthorn plant parts. (from Li and Schroeder 1996, with permission).

Plant part	Possible uses		
Bark	Pharmaceuticals		
	Cosmetics		
Leaves	Pharmaceuticals		
	Cosmetics		
	Tea		
	Animal feeds		
Fruits	Volatile oil	→Pharmaceuticals	
		→Drinks	
		→Food	
	Juice	→Sports drinks	
		→Health drinks	
		→Ternary juice	→Food
			→Beverages
			→Brewery
	Pulp	→Oil	→Pharmaceuticals
			→Cosmetics
		→Residues	→Pigment
			→Animal feeds
Seeds	Oil	→Pharmaceuticals	
		→Cosmetics	
	Residues	→Animal feeds	
Roots	Soil conservation, soil erosion, land reclamation		

Germany. In Saskatchewan, fruit production in shelterbelts ranges from good to excellent with an estimated average of 3.25 kg/tree (6-year-old), with some selected plants producing 5–7 kg annually (Schroeder and Yao 1995). Orchard-type plantations with recommended density of 2500 trees/ha in a male to female ratio of 1:8 provided an estimated yield of 20–25 t/ha in British Columbia. The vitamin C and E contents are as high as 600 and 160 mg/100 g of fruit, respectively (Bernath and Foldesi 1992). The pulp and seeds contain triglyceride oils important for such medicinal values, as control of superoxide dismutase activity in mice, and enhancement of the activity of NK cells in tumor bearing mice (Dai et al. 1987; Chen 1991; Degtyareva et al. 1991).

Sea buckthorn is native to Eurasia. The total area of sea buckthorn in China, Mongolia, and Russia, is approximately 810 000 ha of natural stands and 300 000–500 000 ha planted (Sun 1995). Sea buckthorn was introduced from Russia to the Canadian prairies in the 1930's by Dr. L. Skinner at Morden Research Station, Agriculture and Agri-Food Canada, Morden, Manitoba (Davidson et al. 1994). Originally, plantings were limited to ornamental landscapes in gardens. In 1963, the Shelterbelt Centre of the Prairie Farm Rehabilitation Administration (PFRA) of Agriculture and Agri-Food Canada, Indian Head, Saskatchewan, obtained rooted cuttings and planted them in two hedge rows. These hedges have been used as the sole seed source for propagation of sea buckthorn seedlings planted in shelterbelts throughout the Canadian prairies. This seed source is now known as 'Indian-Summer'. It is one of the hardiest and most adaptable woody plants used in prairie conservation programs (Schroeder 1988) and is well adapted to the Canadian environment. Over one million seedlings have been distributed, and over 250 000 mature fruit-producing plants grow on the prairies with an estimated annual fruit supply from these plantings of 750 000 kg. On the Prairies, sea buckthorn is mainly used for enhancement of wildlife habitat, farmstead protection (Pearson and Rogers 1962), erosion control (Cireasa 1986), and marginal land reclamation (Kluchzynski 1989; Balint et al. 1989; Schroeder 1990). The largest sea buckthorn population in North America is in the Canadian prairie provinces, Saskatchewan and Manitoba, where approximately 100 km of field shelterbelts are planted annually (Schroeder and Yao 1995). Two-year-old seedlings are planted for shelterbelts normally in 1–3 rows, 1–2 m apart within rows and 5 m between rows.

There are several immediate challenges to the establishment of a viable sea buckthorn orchard and processing industry. Production technology utilized in Eastern Europe, Russia, and China is dependent on a supply of low cost labor. Efficient orchard management and harvesting methods need to be further developed to make the industry competitive in a North American setting. A cultivar development program that concentrates on high yield, thornlessness, high nutrient and oil contents and efficient fruit harvesting are underway but availability of these new cultivars on a large scale is several years away. The present supply of sea buckthorn fruit for processing relies on plants that have been established in shelterbelts or in orchards established with cuttings and seedlings.

Sea buckthorn is new in Canada and like any other new crop, it will need considerable research to establish a suitable crop management system to fit its special needs. Investigation of sea buckthorn plantings in Saskatchewan and British Columbia have shown that this species has the potential to be cultivated commercially. However, improvement of cultural techniques, such as pruning and pollination to maximize yields, develop effective disease, insect, pest, and weed controls to minimize loss, and develop mechanical harvesting methods to reduce labour costs, are needed. In addition to plant breeding (see Chapter 3) and production methods (see Chapters 4–8), research on chemical components, post-harvesting handling, processing, and value-added product development are also needed before sea buckthorn can be developed into a sustainable new industry.

Chapter 2. Taxonomy, Natural Distribution, and Botany

Thomas S. C. Li

Agriculture and Agri-Food Canada, Pacific Agri-Food Research Centre
Summerland, British Columbia, Canada V0H 1Z0

Hippophae belongs to the family Elaeagnaceae, which in Russian means 'pasted all over' and refers to its numerous fruits, thickly seated on short pedicels. Rousi (1965) reported on chromosome numbers, finding $2n = 24$ in 32 seed samples of *Hippophae rhamnoides* collected from various locations in Europe and Asia. The classification of genus *Hippophae* has been modified over the years. Originally, it consisted of only one species, *H. rhamnoides*, with 3 subspecies, *rhamnoides*, *salicifolia*, and *tibetana* (Servrttaz 1908). Rousi (1971) re-classified Hippophae, recognizing 3 species based on morphological variations: *H. rhamnoides* L., *H. salicifolia* D. Don, and *H. tibetana* Schlecht. Growth habit of sea buckthorn varied from a tree (H. salicifolia) to a dwarf bush (*H. tibetana*) and a bush-like habit (*H. rhamnoides*). *Hippophae rhamnoides* was further divided into 9 subspecies: *carpatica, caucasica, fluviatilis, gyantsensis, mongolica, rhamnoides, sinensis, turkestanica,* and *yunnanensis* (Rousi 1971). The differences among these subspecies are mainly size, shape, the number of main lateral veins in the leaves, and quantity and color of stellate hairs. Racial divergence does exist, even within each subspecies. At that time, taxonomists were still debating on the precise classification of the genus because of the variation found in the Himalayas and adjacent areas of Central Asia where the primary differentiation of the genus took place (Rousi 1971). In 1978, Liu and He added a fourth species, *H. neurocarpa* S.W. Liu & T.N. He, found on the Qinghai–Xizang plateau of China. Taxonomists in China have modified the classification a few times over the years. Lian and Chen (1997), Lian (1988), and Lian et al. (2000) reported the most recent classification of Genus *Hippophae*. There are a total of 6 species and 12 subspecies:

Hippophae salicifolia D. Don
H. rhamnoides L.
> subsp. *carpatica* Rousi
> subsp. *caucasica* Rousi
> subsp. *fluviatilis* van Soest
> subsp. *mongolica* Rousi
> subsp. *rhamnoides*
> subsp. *sinensis* Rousi
> subsp. *turkestanica* Rousi
> subsp. *yunnanensis* Rousi
H. goniocarpa (Lian) X.L. Chen & K. Sun

subsp. *litangensis* Lian & X.L. Chen
subsp. *goniocarpa* Lian
H. gyantsensis (Rousi) Lian
H. neurocarpa S.W. Liu & T.N. He
subsp. *stellatopilosa* Lian & X.L. Chen
subsp. *neurocarpa* S.W. Liu & T.N. He
H. tibetana Schlecht.

This latest classification was generally validated by Bartish et al. (2002) with modern technology. Their analyses were based on chloroplast DNA (cpDNA), and a combined data set of morphological characters and cpDNA. They suggested that *H. goniocarpa* subsp. *litangensis* should be recognized as a species, since *H. goniocarpa* and *H. litangensis* are clearly not monophyletic, a single lineage of evolution, but rather are sister to two different species in the analysis (Bartish et al. 2002).

Natural sea buckthorn stands are also widespread in Europe on river banks and coastal dunes along the Baltic Coasts of Finland, Poland, and Germany (Rousi 1971; Biswas and Biswas 1980; Kluczynski 1989), along the Gulf of Bothnia in Sweden, and coastal regions of the United Kingdom (Baker 1996). In Asia, the majority of *Hippophae* species are distributed around the northern region of China, throughout the Himalayan region, including India, Nepal, Bhutan, and in the northern parts of Pakistan and Afghanistan (Lu 1992). It grows on hills and hillsides, in valleys and river beds, along coastal regions and on islands, in small isolated or large continuous pure stands, or in mixed stands with other species of shrubs or trees.

Sea buckthorn is a woody, fruit-bearing, deciduous, dioecious, spiny shrub or tree covered with silvery scales. It is usually spinescent and usually reaches 2–4 m in height, although some in China can reach up to 18 m, and others grow no higher than 50 cm. It has brown or black rough bark and a thick greyish-green crown (Bailey and Bailey 1978). Alternate and willow-like leaves emerge 7–10 days after the flower buds, ranging from 3 to 6 cm in length, narrow, and lanceolate with a silver-grey color on the upper side (Synge 1974). Flowers inconspicuous, apetalous, and open before the leaves. Fruit drupe-like and yellow to red at maturity, consisting of a single seed surrounded by a fleshy hypanthium (edible pulp with epidermis). A sea buckthorn plant of fruit-bearing age has 3 types of shoots: vegetative, mixed, and flowering. Vegetative shoots are formed from latent buds on 2–4 year old limbs. Mixed shoots are from buds on 1-year-old wood. Flowering shoots appear at the base of vegetative and mixed shoots. Up to 80% of the fruits are borne on second year wood.

Sea buckthorn tolerates low temperatures to –43°C (Lu 1992), high soil pH up to 8.0, and salt from sea water around the costal regions (Bond 1983). The plant's extensive root system is capable of holding the soil on fragile slopes. It can be planted on marginal land due to its symbiotic association with nitrogen-fixing actinomycetes on its root system (Akkermans et al. 1983; Dobritsa and Novik 1992). Roots of sea buckthorn are also able to transform insoluble organic and mineral matter in the soil into more soluble forms (Lu 1992). The plant rapidly spreads by rhizomatous roots, and suckers will develop within

Fig. 2.1. Diagrammatic representation of flowering and reproductive parts in sea buckthorn (*Hippophae rhamnoides* L).

A, male flowering branches on part of main stem; B, female flowering branches on part of main stem; C, fruiting branch with young leafy shoots; D, male flower, and a bract, showing the inner surface; E, female flower, cut and opened out; F, drupe — partly cut away to show the stone; G, stone – part of the membranous covering cut away to show the seed; H, seed; I,J, upper and lower surfaces of leaf. Perianth green, anthers yellow; bracts densely clothed with reddish-brown scales. Leaves dark blue-green dotted with shining colorless scales, lower surface silvery-gray due to scales. Fruit yellowish-orange.

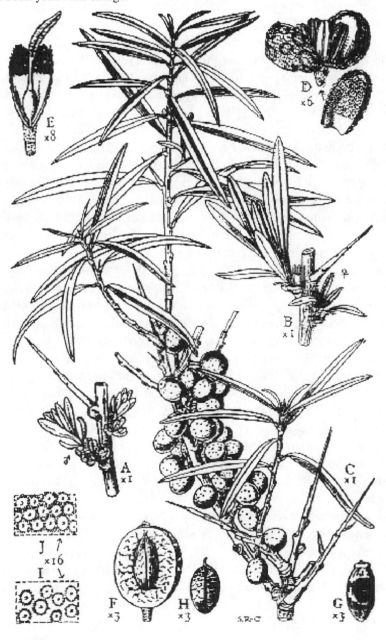

2–3 years of planting, which quickly colonize adjacent areas. The presence of thorns, terminal and lateral, is a biological characteristic of the sea buckthorn. Observations have shown large variation in terms of thorn density, length, shape, and sharpness in natural populations. In evaluating its varieties and selected clones, plants are divided according to this characteristic into slightly (either absent, soft and 1–2 mm long), medium, and heavily thorny groups. In Russia, Mongolia, and Germany, thornless or nearly thornless, and soft-thorn cultivars have been developed recently.

Sea buckthorn is a dioecious species with male and female flowers on separate trees. The sex of seedlings cannot be determined until the setting of flower buds in 3–4 years after seeding (Synge 1974). These are formed mostly on 2-year-old wood (or third leaf) differentiated during the previous growing season (Bernath and Foldesi 1992). Female plants begin to set fruit in 4–6 years from seeds, 2–3 years from cuttings. Male buds are larger than female buds with 6–8 covering scales. The smaller female buds are more elongated and have only two covering scales. The male inflorescence consists of 4–6 apetalous flowers. The female inflorescence usually consists of a single apetalous flower consisting of a pistil, a hypanthium and 2-lobed perianth, and occur in small racemes in the leaf axils (Rousi 1971). The male flower has somewhat longer, oblong perianth leaves and the stamens are all of about equal length. Both male and female flower buds within the same species open at the same time, about 1 week prior to the leaves, in mid to late May. Pollen is released in large quantities when air temperatures reach 6–10°C. Female flowers are receptive for approximately 10 days. Neither the male nor the female flowers produce nectar, which do not attract insects; thus, pollination of female flowers depends entirely on the wind to spread pollen from male flowers.

Sea buckthorn plants may be derived from seeds or cuttings. Seedlings propagated from seeds display variable height, growth form and fruiting characteristics. Clones are vegetatively propagated from cuttings, suckers or tissue culture, normally are produced from a single source and therefore are genetically identical to the parent plant. This method is preferred since the sex of the plant is predetermined and allows control of male–female ratios in the orchard from the initial planting.

Sea buckthorn is a pioneer species, and often is the first woody species colonizing open areas, such as abandoned farmland, wasteland, and rocky islands. It has numerous stems and will form a colony, spread by rhizomes, if left to grow naturally. The root system of sea buckthorn consists of a light-colored, thick, rope-like primary root with horizontal rhizomatous side roots. On sandy meadow soil, the sea buckthorn forms a root system of the surface horizontal type which can be found in a layer 0.1–0.5 m deep. This characteristic is particularly important since sea buckthorn is used to control soil erosion with this rapidly spreading root system.

Like other members of the Elaeagnaceae, 1–2-year old sea buckthorn plants have root nodules, which are small round, and yellowish, containing *Frankia*, a nitrogen-fixing microorganism (Gatner and Gardener 1970). This phenomenon is of considerable ecological significance. Stewart and Pearson (1967) showed that large increases in total nitrogen occurred in a natural stand of sea buckthorn as a result of nitrogen fixation. Total soil nitrogen increased by 1.5 times within 3 years of planting, and by 3–6 times within 50

Fig. 2.2 Sea buckthorn fruits.

years or less of planting in a sand dune ecological system. Although sea buckthorn has the potential to transfer significant amounts of nitrogen to the soil, the level of nitrogen fixing activity varies in response to external factors, such as climate and the nutrient status of the soil (Stewart and Pearson 1967).

Sea buckthorn fruits have attractive colours, varying from yellow through orange to red. Their size varies from 4 to 60 g/100 fruit among genotypes in natural populations, and exceeds 60 g in some Russian cultivars. Average fruit weight for the Canadian cultivar 'Indian-Summer' ranges from 20 to 40 g/100 fruits. Fruit ripening occurs about 100 days after pollination. The fruit is considered ripe when it becomes bright yellow, orange or red in colour in late August to mid-September. Sea buckthorn shows diverse fruit shapes, from flattened spherical, cylindrical, ovate or elliptic, to many irregular shapes. The combination of fruit colour, shape, and size increases the ornamental value of the plant.

Like any other crop, yield of sea buckthorn is affected by factors such as genotype, soil condition, annual precipitation, temperature regime, crop management, number of fruit-bearing branches, and time and methods of harvesting (Kondrashov 1981*a,b*). Yield data for sea buckthorn are scarce since most fruit is collected from natural habitats, soil and water erosion control plantations, and field shelterbelts.

Chapter 3. Plant Breeding

Thomas S. C. Li

Agriculture and Agri-Food Canada, Pacific Agri-Food Research Centre
Summerland, British Columbia, Canada V0H 1Z0

There is a large morphological diversity among sea buckthorn seedlings and mature plants even within each subspecies (Rousi 1971), which is a good indication of excellent opportunities for plant improvement by breeding or selection for desired characteristics for any specific region. Observations of growth and plant size vary by geographical distribution (Yao and Tigerstedt 1994), which indicates that plant selection by geographic area is important and valuable. The first sea buckthorn breeding programs began with mass selection from natural populations. This method is still a common practice (Wahlberg and Jeppsson 1990), although it is gradually being replaced by hybridization (Yao and Tigerstedt 1994; Huang 1995). Polyploid breeding was reported in Russia where autotetraploids were induced by colchicine (Shchapov and Kreimer 1988). There have been no reports of genetic engineering of specific traits in sea buckthorn.

Breeding of sea buckthorn has been conducted for decades in Russia (Goncharov 1995), Ukraine (Gladon et al. 1994), and China (Huang 1995), and recently breeding programs have started in Sweden (Wahlberg and Jeppsson 1990, 1992), Finland (Hirrsalmi 1993; Yao and Tigerstedt 1994), Germany (Albrecht 1993; Müller 1993), and Canada. The Finnish and Swedish programs have concentrated on mass selection and hybridization. In Germany, the objective was to establish a collection of sea buckthorn that represented a wide genetic spectrum (Müller 1993). The Canadian breeding program includes a long-term breeding population made up of progeny from diverse foreign collections and a short-term breeding population of superior individual plants selected from local plantations (Schroeder 1995).

Important characteristics which need improvement in sea buckthorn are yield (Kondrashov 1986a; Huang 1995), fruit size (Buglova 1978), winter hardiness (Kalinina 1987), thornlessness (Hirrsalmi 1993; Albrecht 1993), fruit quality and early maturity (Yao and Tigerstedt 1994), growth habit and long pedicel suitable for mechanical harvesting (Wahlberg and Jeppsson 1990), and nitrogen-fixing ability (Huang 1995). In orchard plantation, to reduce the labour cost, the characteristic of the least amount of, or no suckers produced in sea buckthorn new varieties development has become an important goal (Li, personal oberservation).

Kondrashov (1986a,b) reported that average fruit weight and number of flower buds per unit length of branch are the most reliable traits for yield selection in the Altai region of Siberia. P.L. Goncharov of Russia (personal communication) pointed out that thornlessness could be selected in young seedlings; however, yield of each breeding line cannot be determined until at least the fourth year. Fruit size, thornlessness, and hardiness reportedly are controlled by quantitative genes, and selection for one of these characters will not adversely affect others (Huang 1995).

The amount of constituent substances in the fruit and seed are important factors in breeding and selection programs. The fruits of sea buckthorn are rich in vitamin C, but its content varies considerably between populations in different geographical areas (Li, unpublished data). Karhu and Ulvinen (1999) reported that the correlation between vitamin C concentration and fruit size was low. The vitamin C level in hybrids is related to the level of the particular population of the male parent. In addition to vitamin C, the substances and quantities of particular interest in selection work are riboflavin, niacin, folic acid, tocopherol, flavonoids, carotenoids, and unsaturated fatty acids. Cultivars rich in valuable substances may be preferred in spite of their lower potential yield.

Closely linked to the amount of constituent substances is fruit character. Pigments and aromatic compounds are located primarily in the fruit skin and hypanthium. If the primary purpose is to extract these substances, then fruit size, skin properties, fruit pigment and colour fastness are important. Fruit size is not a major breeding objective in this case as smaller fruits may contain more value than larger fruits. If seed oil extraction is a major goal, seed size and oil content is of critical importance. It is not known if seed shape has any effect on oil extraction, and cultivars with large seeds and high oil content while valuable for processing, may be perceived as 'seedy' in cultivars developed for the fresh market.

Skin toughness is an important character that affects suitability for mechanical harvesting. Skin damage during harvesting is undesirable since it leads to crop losses and lowers fruit quality. Fruit abscission and pedicel length affect the efficiency of mechanical harvest. Fruit with long pedicels and which are easily removed from the plant are desirable. Fruit shape does not appear to be important in mechanical harvesting (Li, personal observation).

The quantity of pollen and time of pollen releasing, which needs to be synchronized with female flower development, directly effects the quantity and stability of yields. One of the objectives in sea buckthorn breeding is to obtain an optimum quantity of pollen in male plants at the same time as the female plants are flowering and receptive (Buglova 1981).

Development and selection of new clones for orchards is being pursued cooperatively by the PFRA Sheltertbelt Centre, Pacific Agri-Food Research Center, and the University of Saskatchewan. Sea buckthorn planted in Saskatchewan shelterbelts are being surveyed and superior cultivars selected from these populations. Selection criteria include yield and fruit quality, skin toughness, thornlessness, and growth habit. Male cultivars with superior pollen production are also being selected.

Limited acreage of *H. rhamnoides* subsp. *sinensis* has been planted in Canada. This subspecies originated in China, where it grows naturally and has been used extensively for soil erosion control and land reclamation. Imported seeds have been tested at Indian Head, Saskatchewan and the Okanagan Valley of British Columbia. Initial observations indicate that it is hardy and grows well. However, it is not well adapted to mechanical harvesting and the characteristics of fruit and seed are less desirable than other species.

There are many clonally propagated cultivars on the market. However, most have been developed in Russia, China, and Europe and have not been adequately tested for their suitability, quality, and yield in Canada. Recommendations of these cultivars for propagation in Canada is not possible until they are properly evaluated.

Chapter 4. Land Preparation, Orchard Design, and Planting

Thomas S. C. Li [1] *and W. R. Schroeder* [2]

[1] *Agriculture and Agri-Food Canada, Pacific Agri-Food Research Centre, Summerland, British Columbia, Canada V0H 1Z0.*
[2] *Agriculture and Agri-Food Canada, P. F. R. A., Shelterbelt Centre, Indian Head, Saskatchewan, Canada S0G 2K2*

Land preparation before planting is important since a sea buckthorn orchard should be planted to last 10–15 years, the expected life span of the crop with maximum productivity. The objectives of site preparation is to eradicate weeds, prepare a planting bed and eliminate soil drainage problems. Ideally, land preparation should begin at least 1 year before planting. Site selection is important since sea buckthorn is a sun loving crop. In its natural habitat, sea buckthorn normally forms a very densely branched bush. Those branches situated in the centre of the bush without sufficient light quickly die back, but remain in position for a long time and form a close thorny lattice (Skogen 1972). The planting site must be cultivated at least twice to remove all the roots of perennial weeds. Previous farming practices or cropping history of the land, such as residual levels of herbicides and pesticides, disease, insect and weed infestations and soil conditions, may affect the growth of sea buckthorn. Site preparation practices vary depending on soil type and condition, previous crop, vegetation cover, and climate of the region. For sandy soils, surface tillage using a cultivator is usually adequate. Organic matter can be increased by adding moderate amounts of well rotted manure or planting a green manure crop for 1 or even 2 years prior to planting, and working it into the soil before seeding. With medium to heavy textured soils (silts and clays) the site should be deep tilled and disced. Where a hard pan exists restricting drainage, subsoiling may be necessary.

Site preparation techniques include:

a) Subsoiling — involves loosening a narrow strip of soil 50–75 cm deep to breakup a hardpan layer.

b) Summerfallow — involves cultivating the site 1 year prior to planting to accumulate soil moisture and control germinating weeds.

c) Cover cropping — involves a cover crop before planting is valuable to increase organic matter in the soil. Barley, oats, winter wheat or fall rye at the rate of 80–150 kg/ha can be seeded in the spring and plowed under before seed set to allow decomposition. Sea buckthorn is planted the following spring. This practice is especially effective on sandy, erodible sites.

d) Discing — fall discing loosens soil providing better reception and retention of moisture as well as better soil tilth for spring planting. Spring discing 2-3 weeks prior to planting controls early germinating weeds.

Sea buckthorn is a hardy plant well adapted to the rigorous climatic conditions in most parts of Canada. The main climatic limitations are extremes of heat, drought or flooding, hail storms and high winds. During the winter, heavy snow loads may damage or break branches. It is important that growers be aware of the rates of occurrence of these incidents at the proposed site. Although sea buckthorn is a drought tolerant species, precipitation during the growing season is an important factor affecting fruit yield, and sea buckthorn orchards should be restricted to areas receiving a minimum of 400 mm of annual precipitation. Areas receiving less than this may require supplemental water to ensure fruit yields are maximized. Most natural populations of sea buckthorn grow in areas receiving annual precipitation of 400–600 mm (Li and Schroeder 1996).

Sea buckthorn can survive summer temperatures up to 40°C; however, Lu (1992) reported that temperature above 30°C burned leaves on newly planted seedlings. In some regions plants ceased to grow under high summer temperatures (Li, personal observation). These observations may be the reason why efforts to introduce sea buckthorn to the plains of Asia and subtropic regions often result in poor growth or even failure. In British Columbia, sea buckthorn showed temporary signs of stress, such as wilting and stunted growth over a period of time, during the peak of summer with temperatures were above 35/25°C (day/night).

Sea buckthorn normally is transplanted (in Canada) or direct-seeded (China) in the spring. Water must be supplied for its establishment. For commercial production in orchard plantations, the site must be selected after considering climate, soil texture and conditions, and available water supplies. Cultural management considerations include land preparation, seed germination, fertilization, spacing, pruning, irrigation and disease and weed control.

In its natural habitat, sea buckthorn is found on slopes, riverbanks, and seashores in a wide range of soil texture (Li and Schroeder 1996). The plant is adapted to a wide variety of soils and it will grow on marginal land, including sandy, gravelly soils with poor nutrient and water retention. On these sites supplementary fertilization, particularly phosphorus at planting time, and irrigation once the shrubs are fruiting may be required. Orchards located on fertile soils rarely require fertilization. Sea buckthorn thrives on well drained light to medium sandy loam. Clay and heavy loam without organic improvements are not suitable. Light sandy soil has low moisture retention capacity and may be improved by the addition of organic matter, manure or various composts. Sea buckthorn will not grow well, perhaps not at all, on compacted wetlands or soils subject to long periods of flooding. Water table depth in the growing area should be greater than 1 m.

Sea buckthorn lives in symbiosis with actinomycetes, order Actinomycetales, which have a low tolerance for acid soil, and prefer a soil pH 5.4–7.0 (Wolf and Wegert 1993). Sea buckthorn prefers soil of pH 6–8. In China, plants have been found in soils ranging from pH 5.5–8.3, although Lu (1992) reported that sea buckthorn thrives best at pH 6–7. Sea buckthorn has moderate tolerance to saline soils. A general rule of thumb is if the site is saline but grows an acceptable crop of barley, sea buckthorn will grow as well, however, its growth and fruit yield will be reduced. Ideally, soil salinity levels should be less than 1.5 mS/cm.

Orchard design (Fig. 4.1 and Fig. 4.2), because of the long-term nature of orchards and the biology of sea buckthorn, requires careful consideration. Such factors as row orientation, plant spacing, and proper distribution of male plants among the females are key considerations. Beldean and Leahu (1985) reported that fruit yields are greatly influenced by sunlight exposure and markedly reduced by shade. Consideration of row planting in a north-south direction is strongly suggested to maximize penetration of sun light to the plants.

Since sea buckthorn is a dioecious shrub, male plants must be sufficiently distributed among the female plants in the orchard to assure adequate pollination and maximize fruit set. On the prairies, sea buckthorn for shelterbelts is grown from seeds and seedlings of unknown sex planted. This practice result in an undesirable distribution of male and female plants and usually results in more male plants than required. For this reason, vegetative propagation from promising mature plants of known sex is preferred. Recommendations for male:female ratio vary. Gakov (1980) considered that 6–7% male trees is adequate for pollination, whereas Albrecht et al. (1984) and Wolf and Wegert (1993) recommended 8–12%. The Siberian Institute of Horticulture in Russia recommended one male:female mixed row for every two rows of female plants. In the mixed row every fifth plant is a male. Goncharov (1995) reported that this design gave significantly higher fruit yields compared to other designs. For effective pollination, the male variety should be cold-resistant, have a long flowering period, provide an adequate amount of pollen, and grow vigorously (Garanovich 1995). These factors should be considered carefully since pollinators are reported to have an appreciable effect on fruit set and size, flavor, and ripening (Buglova 1981).

If seedlings are planted, it will result in an undesirable distribution of male and female plants within each planting. To resolve this problem, once plants have flowered or as soon as the sex of the flower buds can be identified, remove male plants and replace with female plants. It is recommended that nursery rows be planted with seedlings at the same time as orchard planting. Plants in the nursery rows can be used as a source for future replacement of male plants in the orchard. The planting design, illustrated in Fig. 4.1, is recommended for optimum pollination of each female plant in the orchard.

Seedlings are normally harvested from the nursery in the fall and stored in cool storage (2–4°C) over the winter for spring planting. Root pruning before planting is important since part of the root may be mechanically damaged during harvest or become diseased during the storage period. Roots should be cut back to a length of 10–15 cm to encourage vigorous root development. Sea buckthorn roots should be kept shaded and damp and never be allowed to dry during storage and planting. Seedlings and rooted cuttings bene-

Fig 4.1. Recommended sea buckthorn planting design.

```
X O O O X O O O X O O O
O O O O O O O O O O O O
O O O O O O O O O O O O
O O X O O O X O O O X O
O O O O O O O O O O O O
O O O O O O O O O O O O
X O O O X O O O X O O O
O O O O O O O O O O O O
O O O O O O O O O O O O
O O X O O O X O O O X O
O O O O O O O O O O O O
O O O O O O O O O O O O
```

X — Male plants, O — Female plants

Fig. 4.2. Orchard planting.

fit from a pre-planting root soak in water, but this soak should not exceed 4 hours. If fertilizer is needed, it should be well mixed in the soil to avoid direct contact with the root and airholes should be eliminated during planting to prevent the shrub from settling too deep.

Seedlings, rooted cuttings, tissue cultured plants, or suckers can be planted in the dormant stage in the early spring. Planting at this time minimizes stress and takes advantage of new growth after dormancy. If possible, plant trees on a cool cloudy, calm day with minimum wind. Newly planted fields should be irrigated immediately after planting. A short hardening-off period before early spring planting is required for non-dormant plants. The hardening-off process involves allowing the plants to adapt gradually to field conditions by adjusting (lowering) temperature and providing some moisture stress. This process is necessary especially when planting rooted softwood cuttings in late summer or early fall. Planting at this time allows some time for more root growth, but also provides enough time for natural fall hardening.

In orchard plantations, between-row spacing must be sufficient to allow access to the plants with mechanized equipment for mowing, tillage, pest control, pruning and harvesting. To facilitate this, rows should be spaced 4–5 m apart. Wolf and Wegert (1993) recommended a spacing of 1 m within the row and 4–4.5 m between rows to allow equipment access, with rows oriented in a north-south direction to provide maximum light. High density orchards (1 x 1 m) are being considered in Europe to facilitate the use of over-the-row

Table 4.1. Row spacing and number of plants required.

In-row spacing (m)	Between-row spacing (m)	Plants required	
		per hectare	per acre
1	5	2000	735
1.5	5	1333	555
1	6	1666	605
1.5	6	1111	465

harvesting equipment (Olander 1995). In-row spacing needs to be sufficient to allow proper shrub development and 1.0 –1.5 m spacing are recommended. The required number of plants per hectare is based on different plant spacings as shown in Table 4.1. Dense orchard designs allow for higher early yields and optimize land utilization, whereas wider between-row and within-row spacings provide better orchard ventilation and reduce the possibility of spreading disease. Wide between-row spacings make sucker control using mechanized equipment easier. Sucker control is important in orchard plantations to avoid mass formation of unwanted seedlings within and between rows. A within-row spacing of 1.5 m facilitates pruning and mechanical harvesting of individual shrubs. Trees should be planted slightly deeper than the depth at which they grew in the nursery or propagation container. This is especially important for plants in containers as once the rooting medium becomes exposed, the plants dry out very quickly.

Hand planting makes it possible to give maximum individual attention to each planted tree. Planting holes must be large enough to accommodate the root mass. A sharp spade works well, augers can glaze the walls of the hole, inhibiting water movement and the penetration of roots.

The following procedure works well for sea buckthorn:

a) Insert shovel vertically with blade reversed; push handle forward, then pull soil back and out of the hole.

b) Straighten back of hole and insert tree at proper depth.

c) Place tree in proper position, fill hole half way with soil and pack.

d) Fill the hole with loose soil.

If planting in herbicide treated soil, care must be taken not to put treated soil into the planting hole because contact of the roots with herbicide will damage the seedlings.

Mechanical planting operations will require a tractor operator, one or two planters, and a walker to ensure that seedlings have been planted properly and at the correct spacing, as well as to provide a constant supply of seedlings for the planters. Planting should be postponed if soil moisture is below the wilting point or above field capacity. Spacing between seedlings can be regulated by mechanical means (horn attached to measuring device on the planter wheel), by dragging a chain and planting when the last seedling planted passes by the chain end, or by marking one of the planting wheels. Equalized air pressure in packing wheels is important to ensure optimum tracking and planting depth. In heavy soils, the coulter needs to be adjusted and running vertically to the direction of the planter.

Chapter 5. Soil Fertility and Soil Moisture

Thomas S. C. Li

Agriculture and Agri-Food Canada, Pacific Agri-Food Research Centre
Summerland, British Columbia, Canada V0H 1Z0

Sea buckthorn thrives in deep, well drained, sandy loam soil with ample organic matter (Wolf and Wegert 1993) and good growing conditions produce higher yields and better fruit quality (Wahlberg and Jeppsson 1990). In arid or semi-arid areas, water must be supplied for establishment of any sea buckthorn plantation. Information in the literature regarding the cultivation of sea buckthorn is limited. In Saskatchewan, seedlings planted in shelterbelts are often under stress because of lack of soil moisture and nutrition. For commercial production in orchard plantations, cultural management, especially soil fertility and soil moisture is important. A production guide has been published that provides guidance to growers for practical management of sea buckthorn (Li and Schroeder 1999).

Most of the soil fertility research on sea buckthorn was conducted in Russia, which indicated that sea buckthorn, like any other crop, requires adequate soil nutrients for a high yield of good quality fruit, although it requires less N, P, K fertilizers then fruit trees such as apples and pears (Trunov 1996). However, sea buckthorn responds well to phosphorus fertilizer, especially in soils low in phosphorus. Dr. A. Bruvelis of Latvia (personal communication) recommended application of 600–800 kg/ha of calcium superphosphate plowed deeply into the soil. Garanovich (1995) reported that in Belarus, a single winter top-dressing with mineral fertilizer (N:P_2O_5:K_2O; 100:200:100 kg/ha) improved fruit size, yield and quality. Martem'yanov and Khromova (1985) indicated that best growth was obtained by applying peat compost at 60 t/ha and 50 kg/ha each of N, P_2O5, and K_2O. In Siberia, after 5 years total fruit yield increased by 23% when N, P_2O_5, and K_2O at 60:60:60 kg/ha/yr were applied to a black calcareous soil (Predeina 1987). Montpetit and Lalonde (1988) cautioned that nitrogen fertilization can adversely affect root nodulation and it delays the development of nodules after inoculation with *Frankia*. Similar results have been shown for other nitrogen-fixing woody plants (Mackay et al. 1987). Mishulina (1976) reported that foliar sprays with micronutrients, Cu, Mo, Mn, I, B, Co, and Zn, increased fruit weight by up to 34.5%. In China, application of 100–150 t/ha of compost or 400–500 t/ha of green manure is recommended prior to planting (Lu 1992). Wolf and Wegert (1993) and Li and Schroeder (1999) noted that precise fertilizer recommendations for each orchard should be determined based on the results of leaf and soil analysis.

Soil analysis is the most accurate and reliable guide to fertilizer and lime requirements. It is important to determine soil fertility and pH levels before planting, so that necessary lime and fertilizer can be applied to soil (Li and Schroeder 1999). Soil acidity and alkalinity, except at extreme levels are not limiting factors for sea buckthorn plantations, however, the most favorable range is pH 5.5–7.0. If the soil pH is too low, it can be corrected by application of lime such as dolomitic limestone. Use of some dolomitic limestone is recommended since it contains a significant quantity of magnesium which is an essential, and often deficient, plant nutrient. Quicklime, caustic lime, and burned lime are not recommended on agricultural land.

Increased production requires proper levels of nutrients to ensure flower bud development and good fruit size and quality and flower bud development. A good feeding program in mature trees is as important as careful attention to pruning procedures. Requirement of various nutrients can be determined by taking leaf samples starting at the early flowering stage. The yield depends on the variety, planting spacing, tree condition and age, as well as on the grower's management and the nutritional feeding program applied. Sea buckthorn is capable of fixing nitrogen by its root, however, a small amount of nitrogen or larger amount of slow released nitrogen applied just after planting is beneficial (Li, unpublished data).

There are various methods of fertilizer application. It can be broadcasted on the soil surface and incorporated into the soil with tillage. The top-dressing method can be used when sea buckthorn is growing. Fertigation is also effective.

Manure or compost supplies plant food over a period of time and cow and poultry manures are commonly used. Maximum application rates of dairy manure should be about 45 t/ha and poultry manure should be applied at no more than 20 t/ha on cropped land. A cover crop before the land is to be planted with sea buckthorn is valuable in increasing organic matter in the soil and preventing nutrient losses and erosion by wind and water. Barley, oats or winter cereals such as winter wheat and fall rye can be seeded (80–150 kg/ha) in the fall and plowed under in the early spring to allow decomposition before sea buckthorn planting.

There is very limited information in the literature regarding the effects of nutrient deficiencies on sea buckthorn. However, based on observations over the years, the following physiological disorders were evident, and these are commonly caused by nutrient deficiency.

Major Elements

Foliage is pale green or yellowish and later leaf senescence and dehiscence are accelerated, plants are smaller than normal, leaf area is reduced. To correct this disorder, nitrogen and phosphorus fertilizer should be applied early in the spring in the form of ammonium nitrate at 20 g/m^2 and phosphate fertilizer (superphosphate) should be added in the fall at a rate of 20–30 g/m^2 (these rates are based on the results of soil analysis). Pale leaf colour, marginal chlorosis, scorch, shortening of stem internodes, and death of the terminal bud are normally caused by potassium deficiency. Potassium fertilizer should be added with the phosphorus at 20–25 g/m^2 as potassium chloride.

Minor Elements

Terminal leaves are normal, basal leaves show marginal chlorosis with a V pattern, this symptom occurs when the magnesium level is low and the potassium is high or soil is acidic. This is especially noticeable in young, vigorous trees where defoliation begins from the base shoots and progressively affects the leaves above. Zinc deficiency causes delayed opening of flower and leaf buds in the spring, small, chlorotic leaves, shortened internodes and small leaves along the shoot also reduced growth and produced small fruit. Iron deficiency can result when there is insufficient iron in the soil, sufficient but unavailable iron, or sufficient and available iron that is not properly utilized in the plant. The symptoms included loss of chlorophyll and leaves become chlorotic, and interveinal areas become yellow but the veins remain green. Symptoms develop first on young leaves.

Smaller Crowns and Lower Yields are the Result of a Shading Effect

Sea buckthorn can only be grown in well-lit, unshaded areas. Proper prunning every year in the winter will improve yield. Too much water or high ground water level will cause poor growth and root rot. Planting sea buckthorn on slight slopes or sandy loam soil with good drainage is preferred. Lack of adequate soil moisture may cause small leaf area and low fruit set. Winter or frost damage will cause die back of branches with late emerging small leaves. Lack of soil moisture will result in wilt of the leaves that turn yellow and drop.

Buglova (1978) reported that weather conditions, especially precipitation, could affect fruit weight. Sea buckthorn is sensitive to severe deficits of soil moisture, especially in the spring when plants are flowering and young fruit are beginning to develop. Under extended drought situations (i.e., >30 centibars), young fruit may dehydrate or abscise (Lu 1992). Sea buckthorn is a moisture loving crop which needs to be irrigated regularly especially in the summer. Garanovich (1995) reported that irrigation was necessary during dry growing seasons. In an unconfirmed report, Lu (1992) indicated that the minimum soil moisture levels needed to cultivate sea buckthorn in medium clay loam, heavy clay, slightly sandy soil and sandy loam were 70, 80, 60, and 65–70%, respectively. Li (1990) reported that crown diameter and fruit yield increased 56 and 47%, respectively, in an irrigated plot with more that 70% soil moisture compared with a non-irrigated plot with soil moisture of 50–60%. These reports suggest that irrigation is beneficial or required during periods of extended drought, especially during flowering and fruit development stages. Surface water, whether from a river, lake, reservoir or farm dugout is usually superior to ground water since it is warmer and contains fewer soluble solids, especially salts. If using well water, the salinity levels should be less than 1 mS/cm. The optimal soil moisture for sea buckthorn is around 70%, inadequate soil moisture causes a reduction of leaf area and fruit size.

Chapter 6. Propagation

Thomas S.C. Li[1] and W.R. Schroeder[2]

[1] *Agriculture and Agri-Food Canada, Pacific Agri-Food Research Centre, Summerland, British Columbia, Canada V0H 1Z0.*
[2] *Agriculture and Agri-Food Canada, P. F. R. A., Shelterbelt Centre, Indian Head, Saskatchewan, Canada S0G 2K2*

Sea buckthorn may be propagated by seeds, softwood and hardwood cuttings, layering, suckers or tissue culture (Table 6.1). Micropropagation using meristem culture has been investigated (Burdasov and Sviridenko 1988; Montpetit and Lalonde 1988) in the past but this technology is only used occasionally in a few nurseries. The easiest and the most successful method of propagating sea buckthorn plants is from seed. Seed propagation is well adapted to producing plants for conservation plantings, which are not intended to be intensively managed. Another propagation method is vegetative propagation, which includes softwood and hardwood cuttings. Polikarpova et al. (1999) reported that sea buckthorn can be successfully propagated by semi-hardwood and fully lignified leafy cuttings taken late in the summer. Traditional methods of establishing sea buckthorn stands involves nursery plants grown from lignified cuttings; however, Kondrashov (1994) reported a new technique involving direct planting of long (40–45 cm) cuttings collected from annual shoots and planted (35–40 cm deep) vertically or at an angle. Reported survival rate for this method was 73–100%. Vegetative propagation methods are the most appropriate for plants utilized in commercial orchards. Vegetatively propagated plants only have a single parent, therefore are genetically identical to the parent plant allowing pre-determination of growth character and sex of the plant. This is very important as indicated in Chapter 4 relative to orchard design.

Seed Propagation

Seed propagation is relatively simple which produces a large number of seedlings at fairly low cost compared with other propagation methods. This method is commonly used when growing seedlings for wildlife habitat, soil reclamation or erosion control. Because the *Hippophae* is dioecious, seed propagation will produce both male and female plants in equal quantities and their flower buds are indistinguishable for sex for about 4–5 years after planting. Also, there is considerable variation among seedlings propagated from seeds. Consequently seed propagation is not the preferred method of propagating plants for commercial fruit production.

Table 6.1. Propagation methods applied to sea buckthorn.

Propagation Method	Comments
Seed	Inexpensive and simple procedure. Seeds contain a mix of genetic material from two parents, consequently seedlings are not identical to their parents. It is impossible to distinguish between male and female plants from seed propagated plants until they start to flower about 4–5 years after germination. Also, some desirable characteristics may not be expressed in seed propagated plants.
Softwood cuttings	Inexpensive, simple, highly successful procedure, but requires some initial capital expenditure. May fruit earlier than sexually propagated plants. Produces genetically uniform plants of known sex. Initially, availability of propagation material may be limited.
Hardwood cuttings	Inexpensive, simple procedure, but rooting success rate is not as high as softwood cuttings. May fruit earlier than sexually propagated plants. Produces genetically uniform plants with known sex. Initially, availability of propagation material may be limited.
Suckers	Simple, inexpensive technique but available material may be limited. Suckers have poor root mass and may be susceptible to transplant shock.
Tissue culture	Complex and techniques are not well developed at this time. Produces genetically uniform plants of known sex. Potentially quick production of large numbers of plants.

For seed extraction, sea buckthorn fruits can be collected from late August to mid winter. Fifty kilograms of fruit will yield 4–5 kg of cleaned seed (30 000–40 000 seeds/kg). Seeds are best extracted from fresh or frozen ripe fruit. Seed is separated by maceration and floating off the pulp and juice. Seeds are surface dried after extraction. For small quantities, seed can be extracted by using a blender set at low speed and with dull blades or the blades wrapped with masking tape. Alternatively, a commercial mechanical finisher machine can be used. Clean and dried (6% moisture) seeds can be stored in sealed containers up to −18°C.

Many factors affect seed germination, seed source (Fefelov and Eliseev 1986), maturity (Eliseev and Mishulina 1972; Vernik and Zhapakova 1986), dormancy (Siabaugh 1974), pre-seeding treatments (Li and Shroeder 1996; Lu 1992), planting media (Salo 1991), seeding depth (Li and Wardle 1999), and soil and environmental conditions (Vernik and Zhapakova 1986). Pre-seeding treatment is important to break dormancy in sea buck-thorn. In a recent study, Li and Wardle (1999) reported that seeds soaked in water or potas-sium nitrate solution (0.01 M) at room temperature (20°C) for 48 h emerged in higher percentages compared to other pre-seeding treatments including soaking in water at 70°C, cooling to room temperature and soaking in gibberellic acid (GA^3) solution (500 mg/L). Seeds will germinate within 3–10 days after the treatments. Fungal problems can be min-imized if the seed is surface-sterilized in a solution of household bleach (1: 9, bleach: water) for 5 minutes.

Storage affects seed viability. Smirnova and Tikhomirova (1980) reported that seeds of *H. rhamnoides* lost 60% of their viability after 4–5 years of storage. Prior to seeding, the seeds should be soaked in water for 48 h and floating seeds discarded. Lu (1992) reported that seeds placed in water for 48 h initially at 70°C gave satisfactory germination rates. Other reports indicated that soaking seeds for 48 h at a temperature of 70°C and stir-ring intermittently until the temperature drops to 10–15 °C may improve the germination rate. The results from our experiment indicated that water temperature did not signifi-cantly affect germination rates, and seed soaking shortened the number of days required to complete emergence.

Internal seed dormancy can be broken by stratification in moist sand for 90 days at 5°C (Siabaugh 1974). At 10–12°C, Lu (1992) reported 13.2% germination after 47 days, compared to 95% in 6 days for seeds at 24–26°C. Vernik and Zhapakova (1986) also reported more rapid germination at 25–27°C than at 20°C. After soaking, seeds should be air-dried before seeding. This technique is useful when sowing outdoors in spring or indoors in a greenhouse. At the Shelterbelt Centre in Saskatchewan, non-stratified seeds sown in late September at a depth of 1 cm and a rate of 100 seeds /m in rows 60 cm apart emerged the following spring with a 90% germination rate.

Seeding depth has a significant effect on seed emergence rate. Li and Wardle (1999) indicated that seeding on the soil surface had higher emergence rates than at depths of 1 and 2 cm. Soil should be irrigated periodically to prevent drying. If seeding in the late spring, seeds should be covered with very light layer of soil. Seeds should start to germi-nate within 5–10 days based on the condition of the seeds and the species, subspecies, or varieties of sea buckthorn.

The stratified sea buckthorn seeds can be seeded indoors in pots in January or early February in a vermiculite/peat moss mixture (40:60). The containers are placed in a green-house with a 16 h photoperiod at a temperature range between 25–27°C and 70–90% rel-ative humidity. Germination will occur in 3–10 days. Immediately following germination and prior to the formation of true leaves, an application of fungicide may be necessary to control seedling damping-off.

Maximum growth may be obtained by using bright full-spectrum flourescent or high-pressure sodium lighting. A soluble starter fertilizer (10–52–10, N–P–K) is applied with each irrigation for the first 3 weeks following planting. After this, a complete soluble fer-tilizer (20–20–20, N–P–K) is applied in the same manner. A moderate amount of air movement is necessary to provide adequate ventilation. Seedlings grow rapidly and will need to be acclimated before transplanting to the field in the spring.

One seedling per pot is allowed to grow for 3 months before transplantation to the field in early May. On light sandy soil, the root is buried 6–8 cm deep to encourage the development of another tier of roots, and newly planted seedlings should be watered as required. In orchard plantings, a spacing of 1 m within the row and 4 m between rows is recommended, although high density planting of 1 m x 1 m is being considered in Europe (Olander 1995).

In outdoor nursery practice, stratified seeds are planted 1 cm deep at the rate of 150 seeds/m2 in the spring. Non-stratified seed should be sown in late September. Seedlings will thrive on medium textured soils at pH 7.0–8.0 and organic matter content of 3–4%. Herbicide should be applied prior to sowing for weed control. Linuron applied to dormant seedlings controls most winter annuals. Defoliated and well nodulated seedlings are lifted after 1 or 2 growing seasons when they have reached the standard height (>25 cm) and root collar diameter (>4.0 mm). Seedlings must be graded, sorted and placed in storage as quickly as possible following lifting. Roots should be stored at –2°C in poly bags to avoid desiccation. Seedlings can also be stored in shipping bundles with no reduction in quality.

Vegetative Propagation

Sea buckthorn can be propagated from both hardwood and softwood cuttings with two main advantages. First, genetic characteristics generally reproduce exactly because the cutting is a clone of the mother plant. Second, cuttings can often flower and produce fruit in less time than a seedling.

Softwood cutting

The maturity of the shoot and physiological condition of the cuttings are important for root formation. Softwood cuttings are collected from selected plants in late spring when shoots are no longer growing but prior to lignification. This is usually the end of June to the beginning of July. Best results occur when cuttings are collected from terminal or side shoots. Cutting length should be 10–15 cm. The cutting usually includes more than two nodes, and the foliage is removed from the lower portion of the cutting. Collect cuttings, preferably on cool dull days in the morning or late afternoon. Cuttings are placed in poly or moist burlap sacks immediately after cutting. The key to success is to keep the Sea buckthorn cuttings turgid during the period from collection until they are well rooted in the propagation facility. Prior to sticking, the cuttings are recut and sufficient leaves are removed so that the lower leaves are not in contact with the medium after sticking. Cuttings should be planted so that at least two leaf nodes are below the surface of the media. Rooting hormone is applied by dipping the base of the stem in water and then in rooting hormone (indole butyric acid, IBA at 0.1%).

The advantage of softwood cutting propagation is its high success rate. Softwood cuttings (10–15 cm long) are taken when shoots begin to become woody. The lower leaves are removed, leaving 2–4 leaves at the tip and the lower end is dipped into rooting hormone before rooting in media, such as sand or perlite. Media should be kept moist at all times. Rooted cuttings should be planted in pots for 1–2 months before transplanting to the field. Avdeev (1976) reported that softwood cuttings collected from an 8-year-old tree in early spring, treated with IBA (50 mg/L) solution and planted in a peat and sand (2:1) mixture under mist, had 96–100% rooting in 9–11 days. Balabushka (1990) compared the effects of indole acetic acid (IAA), IBA and chlorophenoxyacetic acid on rooting of sea buckthorn, and found that IAA root dips were superior to the other hormone treatments. Ivanicka (1988) reported that with or without IBA (0.1–0.3%) treatment, semi-lignified *H. rhamnoides* cuttings rooted readily in a peat, polystyrene granule, and sand (1–2 : 1 : 0.5) mixture under mist in a plastic house. Timing of cutting collection is important. Kniga (1989) reported that in the Kiev region of Ukraine, the optimum time was late May. Kondrashov and Kuimov (1987) reported that cuttings taken in late June from severely pruned branches (pruning was conducted in early spring before bud break), successfully rooted (95–98%) in the greenhouse under mist.

Cuttings require 80–90% humidity to remain turgid during the rooting process. Humidity can be controlled in a number of ways, but the most reliable is with automatic misting units, such as brass mist nozzles controlled by electronic time clocks and a solenoid. The mist system is on day and night for the first 4 weeks. Initially, 10 s of mist is supplied every 30 minutes. Gradually reduce the amount of mist after roots begin to form, until misting is eliminated. It will take 4–6 weeks for the cuttings to root.

Propagation beds are shaded and include mist. The mist bed can be a clear polyethylene tunnel (or mesh shade cloth), located on 7–10 cm of coarse gravel. Basal heat is provided in the propagation bed using a heating cable. Ideally, temperature of the substrate should be 25°C. Adequate ventilation is required. Temperatures in the rooting tunnel should not exceed 40°C. Cuttings can be rooted in any substrate providing good air/water relationships. Media mixes, such as # 2 or # 4 Sunshine Mix (Fisons), Premier Promix 'HP' or a 50:25:25 by volume peat:vermiculite:perlite mixture, have provided good results.

To ensure vigorous plant growth, fertilize the root zone with a weekly application of 20–20–20 fertilizer at 2000 ppm (2 g/L) starting as soon as the first roots are formed and continuing throughout the growing season. Preparing the cuttings early in the summer, thereby allowing time for their establishment before winter, also helps the plants to over-winter successfully.

Hardwood cuttings

Successful rooting percentages from hardwood cuttings varies. Avdeev (1984) reported 86–100% success. Garanovich (1984) reported that in the greenhouse under artificial mist, rooting success was 20% lower with hardwood cuttings than with softwood cuttings, but plants from hardwood cuttings attained heights of 90 cm by the end of the first growing season and could be planted out the next spring, whereas plants from softwood cuttings needed 1–2 years before transplanting to the field. Kondrashov and Kuimov (1987) reported that hardwood cuttings with apices removed were rooted successfully outdoors under plastic in pure sand or a sand and peat (1 : 1) mixture. In a separate experiment, they showed that 2-year-old wood, cut before bud break, and stored for 10 days in sawdust at 10–15°C, gave 100% rooting in the field.

Kuznetsov (1985) recommended taking cuttings before bud break, soaking in water (18–20°C) for 7 days and planting in the field with dark polyethylene mulch. In British Columbia, we obtained 90% rooting of cuttings taken in mid-March, stored in plastic bags at 0°C until May and placed in pots filled with peat in a heated propagation box (18–22°C) indoors under fluorescent light. Lu (1992) reported that hardwood cuttings have not been used widely in nurseries.

Sharp knives or pruning shears should be used for harvesting the 1-year-old shoots. Using pneumatic pruning shears doubles production. To avoid drying during harvesting and transport to the storage facility, cuttings should be covered with damp burlap. The shoots removed from the parent plants are bundled and labeled to show the name of the variety and stored around –2°C until processed for rooting.

One of the main factors determining the quality of the shoots and the rooting ability of the cuttings is the age of the parent plant. The growing period of shoots on older Sea buckthorn trees is shorter than for young trees and the shoots are usually shorter (7–20 cm) with short internodes. The cuttings from older trees, grown in natural habitat or shelterbelts without proper annual pruning, are less likely to root. On parent plants up to 5 years old, 1-year-old whips can reach a length of 30–100 cm depending on the cultivar. Normally 1–3 cuttings, 15–20 cm in length, can be obtained from each whip.

For rooting hardwood cuttings, 1-year-old whips are collected during the winter and stored whole or cut up (15 cm long). Make the lower cut under the bud and the upper one above the bud. Use pruning shears or cut mechanically. The main factor in cutting is that the blade must be sharp for the clean cut. Rough cuts do not heal quickly and often decay. Discard the non-lignified part of the whip and divide the lignified portion into cuttings 15 cm long. Place the cuttings in bundles of 100, dip the cut ends in paraffin wax, affix a label bearing the name of the cultivar and store the bundles at –4°C or in a snow cache. If a snow cache is used dig a trench in the middle of the cache and stack the bundles of cuttings in several rows and layers with the tips up. Cover each layer and row with snow 10–15 cm deep and a 1 cm layer of sawdust. Cover the cache with straw or sawdust and make sure that the snow does not melt while the cuttings are in storage. Before planting, soak the cuttings in water at a temperature of 18–25°C, leaving 2–3 buds exposed, for 5–7 days until the buds swell. Water must be changed daily to prevent stagnation.

These cuttings are then planted in a nursery field. The field should be light soil, rich in organic nutrients with a pH 6.5–7.5. and protected by shelterbelts, with plenty of sunshine. During the year of planting, spring preparation of the field involves working the soil with harrows, discing to a depth of 5–8 cm, and packing. Soil should be irrigated 3–5 days before the cuttings are planted, if the soil moisture is below 70% field capacity.

The best time for planting sea buckthorn cuttings is the mid to late spring. Osipov (1983) reported that 15–20 cm hardwood cuttings planted in spring gave 95% rooting and under proper management 80% of them attained standard size by the autumn. Autumn planting of sea buckthorn cuttings usually gives poor results. Cuttings are planted in a prepared field in twin rows of 70 x 20 cm at a 5–7 cm spacing in the row, and at the rate of 360–400 thousand cuttings per hectare. When the young plants are to be grown in small quantities, the cuttings can be planted in raised beds 1 m wide in a 10 x 10 cm grid. The cuttings are planted vertically. They should be planted out so that 2–3 well-developed buds are visible above the surface. The soil around the cuttings is thoroughly compacted after planting and then irrigated immediately. By the end of the first growing season most of the young plants are ready for transplanting to the orchard, while a few may require another year of growth.

Soil moisture should be monitored throughout the growing season and maintained at 80–100% of field capacity while the cuttings are forming roots. This can be achieved by irrigating once a day during dry sunny weather. In overcast weather the frequency of irrigation can be reduced. After the cuttings have rooted, the soil moisture content can be maintained at 70–80% of field capacity. When managed in this way a rooting and survival rate of 75–90% at the end of the first growing season is possible.

Hardwood cuttings can also be rooted in the greenhouse. Cutting should be chosen from healthy, well developed plants at the fruiting stage in the sea buckthorn orchard or plantation. The cuttings can be stored at 1°C in sealed plastic bags for up to1 month. Cuttings are planted in containers filled with peat moss and perlite mix (3:1, v:v). The pH of the media should be adjusted so that it is greater than 6.5. When planting, leave 2 buds above the media surface. The media is kept moist but not saturated. For optimum rooting, greenhouse temperature is maintained at 5°C during the night and 25°C during the day. Once shoots are growing, the photoperiod is set at 16-hour light and 8-hour dark. When shoots are 2.5–3 cm long, remove the shortest shoot. Rooting normally occurs within 2 weeks of planting. When roots are 1–2 cm long, the plants are fertilized with a soluble fertilizer (20–20–20, N–P–K) once a week. Rooted cuttings are grown in the greenhouse for 6–8 weeks before transplantation in the field. After transplanting, the rooted cuttings need to be irrigated periodically. Rooted cuttings can be directly planted outdoors in the field, but planted in pots under a controlled environment for 1–2 months before transplanting will give better results.

Overwintering cuttings during the first winter sometimes presents a problem especially in a cold environment such as the Canadian prairies regardless of whether the cuttings have been transplanted in the field or stored indoors. They must be acclimated and

Table 6.2. Formulations of plant tissue culture media used for sea buckthorn (Yao 1994).

Compound	Woody Plant Media (mg/L)	Modified Media (mg/L)
NH_4NO_3	400	400
KNO_3		100
$Ca(NO_3)\,4H_2O$	556	30
$MgSO_4\,7H_2O$	370	450
Na_2SO_4		100
$CaCl_2\,2H_2O$	96	
KH_2PO_4	170	200
KCl		50
K_2SO_4	990	
$FeSO_4\,7H_2O$	27.8	27.9
Na_2EDTA	37.3	37.3
$MnSO_4\,H_2O$	22.3	12.8
$ZnSO_4\,7H_2O$	8.6	3.2
H_3BO_3	6.2	3.2
KI		0.5
$Na_2MoO_4\,2H_2O$	0.25	1.2
$CuSO_4\,5H_2O$	0.25	0.5
$CoCl\,6H_2O$		0.1
Myo-inositol	100	100
Nicotinic acid	0.5	0.5
Pyridoxine HCl	0.5	0.2
Thiamine HCl	1	0.5
Glycine	2	
Saccharose		20000
Agar		3750

hardened off prior to planting from the propagation bed. The best method is to overwinter the rooted cuttings in the propagation bed (covered with microfoam or mulched with peat moss or wood chips) or remove them and store in poly bags at –2°C. Alternatively, the cuttings may be transplanted to the field in late August when roots are well developed. This method works well if cuttings have been rooted directly in containers. Transplants should be kept moderately well irrigated until late summer. Weeds should be controlled but deep cultivation close to the plants must be avoided.

Root cuttings

Root cuttings can also be an effective propagation method for sea buckthorn (Dr. A. Bruvlis, personal communication). Root cuttings were planted in pots and placed in a greenhouse for 6 weeks before being transplanted in the field at a spacing of 8 – 20 cm.

Cuttings need to be acclimated to field conditions prior to planting by placing pots in a shady area for 1 week. The best results were obtained in sandy loam at pH 6–6.5 with a medium humus content. Sea buckthorn readily produces suckers within a few years of planting. This is a good source of material for propagation (Kondrashov and Kuimov 1987). The probability of invasion by suckers into surrounding areas is high, routine cultivation and herbicide application are the best control measures for this weediness characteristic of sea buckthorn.

Sucker

A sucker is a shoot that arises on a plant from below ground (rhizomes, which are underground stems). Propagation using suckers is a form of dividing the plant. Suckers are dug out and cut from the parent plant. In some cases part of the old root may be retained, although most new roots arise from the base of the sucker. It is important to dig the sucker out rather than pull it, to avoid injury to its base. Suckers are usually removed with the aid of a pruning knife and/or a shovel. Try to obtain as large a root mass as possible, and do not allow the root mass to dry out prior to transplanting. Suckers are best dug in early spring while still dormant.

Tissue culture

Sea buckthorn can also be propagated by tissue culture with shoot tips, callus, anthers and pollen, cells and protoplasts in culture on defined and semi-defined media. All offspring from a single mother plant will be identical and are members of a clone group. This means that their genetic makeup is identical to the original plant. Laboratory propagation permits year round production scheduling. Tissue culture plants are grown in "season controlled" growth chambers between 25–27°C for optimal growth. Shoot tips can be collected from actively growing branches in June/July, or all year round from greenhouse grown plants. Modified media (Table 6.2) supplemented with 0.035–0.04 mg/L NAA (naphthaleneacetic acid) and 0.3 mg/L kinetin for initiation and multiplication and 0.002 mg/L NAA and 0.1 mg/L kinetin for rooting of regenerants. For initiation and multiplication, 25–27°C under a light intensity of 4000 lux at a photoperiod of 16 h is applied. Acclimation period is 10–15 days where relative humidity is gradually reduced from 90 to 50% (Yao 1994).

One important way of accelerating the plant breeding process is rapid propagation of elite hybrid forms. Technical protocols for culturing isolated tissues and organs for the propagation of superior lines of sea buckthorn are currently being developed. These techniques will be used in plant breeding for the accelerated propagation of new varieties.

Chapter 7. Pruning

Thomas S.C. Li

Agriculture and Agri-Food Canada, Pacific Agri-Food Research Centre
Summerland, British Columbia, Canada V0H 1Z0

Sea buckthorn is a new crop in North America, currently being adopted for orchard planting. Information and experience related to pruning sea buckthorn is very limited. An approach to orchard pruning requires an understanding of crop management techniques, as well as variations due to geographic location, climate, cultivars, and tree spacing. The proper application of specific types of pruning cuts requires an understanding of the following factors.

Specific objectives of pruning sea buckthorn are:

a) to produce and maintain the proper plant size, shape, and architecture.

b) to improve branching habit and strength.

c) to produce and maintain the optimum number of new and young fruiting branches and to remove old and weak fruit bearing branches that are only marginally productive with poor quality fruit.

d) to maintain or increase plant vigor by removing broken, diseased, or insect-infested branches.

e) to increase light penetration thus equalizing the fruiting potential throughout the tree canopy.

f) to induce and maintain an annual bearing habit, to maintain high and predictable yield, and to renew structural units of declining fruitfulness.

g) To improve insect and disease control by ensuring effective penetration of pesticides.

Albrecht et al. (1984) reported that the purpose of pruning sea buckthorn is to train branches and promote growth to facilitate harvesting. It is important to remember that all portions of the tree canopy must receive adequate sunlight for the initiation of flowers, for fruit set and for fruit size with good colouration and even maturity. It was reported that moderate pruning will increase the yield and fruiting life of the plants (Savkin and Mukhamadiev 1983). Sea buckthorn grows up to 2–3 m in 4 years and forms its crown at the base of the main trunk. The crown should be pruned annually to remove overlapping branches, and long branches should be headed to encourage development of lateral shoots. In about the fifth year, the main trunk stops growing, and branches begin to grow from lateral buds. Mature, fruiting plants should be pruned to allow more light penetration if the bush is dense. Sea buckthorn grown under full sunlight produced more fruit than when grown under partial shade (Beldean and Leahu 1985), and proper annual pruning can provide more even exposure of branches to the sun. To protect sea buckthorn from premature senescence, 3-year-old branches should be pruned for rejuvenation (Lu 1992). In Russia, pruning trials were carried out with the goal of creating a hedge to allow efficient mechanical harvesting (Savkin and Mukhamadiev 1983). Similar work is underway in Germany (Gaetke and Triquart 1992) and Sweden (S. Olander, personal communication). However, this technique is not ideal, since the mechanical harvester tended to cut off fruiting branches and these trees will not have any fruit in the coming year.

Vegetative growth is necessary for any plant to maintain its vigor, to provide leaf area, and to develop fruit bearing branches. However, vegetative growth of any fruiting tree is directly competing with fruit formation for the nutrients within the tree, therefore, the development and maintenance of excessive, unproductive vegetative branches should be restricted. Pruning is an excellent method of manipulating the ratio of vegetative growth and fruiting by minimizing unproductive shoot growth and optimizing fruiting. Proper pruning will expose more leaf area to sunlight, which is one of the most influential factors for good growth and higher yield (Beldean and Leahu 1985). A plant that is pruned on an annual basis appears to have increased vigor because it creates new and more productive shoots where the pruning cuts are made.

Tree size has an important influence on the ratio between vegetative and reproductive growth. Large trees tend to produce less fruit per unit of vegetative growth than small trees. Tree size determines the percentage of the total leaf surface that receives adequate sunlight. The earlier regular pruning provides the greater control of tree size. As shown in Fig. 7.1, reducing tree size from 5 to 2.5 m reduces the shaded interior from 24.4% of the tree volume to 1.6%. Excessive nitrogen fertilization will also induce vigorous vegetative growth. High vigor may increase the yield of fruit up to a point, but eventually shoot growth becomes directly competitive with fruit set and results in a reduction of yield.

Pruning should be started the year trees are planted. Sea buckthorn buds break in the early spring, therefore, late winter pruning prior to bud break is preferred. One of the

Fig. 7.1. Effect of tree size on light exposure.

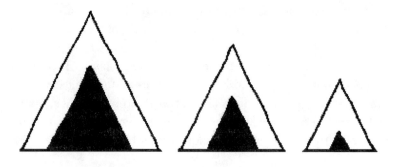

important reasons to prune sea buckthorn early is that spring pruning may cause sap to run down along the pruning cuts which may promote disease later in the season. Summer or late fall pruning is generally not recommended. There are a number of pruning tools available. Secateurs can be used for small branches up to 1.5 cm in diameter. For larger branches, pruning saws or loppers should be used. Large orchards will benefit from the purchase of pneumatic pruners which will save time and make the task easier. Regardless of types of pruning cuts, location, and size, all pruning cuts should be flush and smooth which requires sharp cuts with proper angle to avoid retarding the healing process and to reduce susceptibility to diseases. Wound dressings are not necessary except for large pruning cuts.

There are 4 major pruning cuts:

a) Heading-back cut

This pruning method primarily removes terminal buds and disrupts the natural growth pattern. The disadvantage is that head-back will stimulate total branch and shoot growth. As shown in Fig. 7.2, when heading-back cuts are made on the 1-year-old branches, this usually results in the development of very vigorous shoots from the three or four buds immediately below the cut. These shoots tend to develop with narrow crotch angles and to grow strongly upright. These branches will create undesirable shade to other branches and have little fruiting potential.

b) Thinning-out cut

This pruning cut refers to the removal of an entire shoot or branch (Fig. 7.3). Vigorous shoot growth may develop in the immediate vicinity of the pruning cut, but the effect on adjacent parts of the tree is minimal and the ratio of terminal to lateral buds is largely

Fig. 7.2. Responses of branch growth to heading-back cuts.

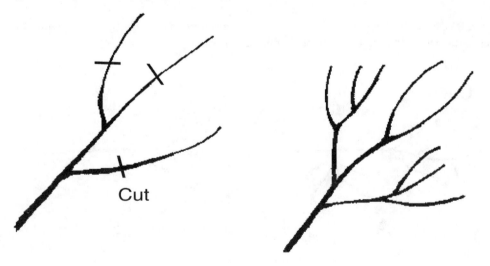

Cut

Fig. 7.3. Responses of tree growth to thinning-out cut.

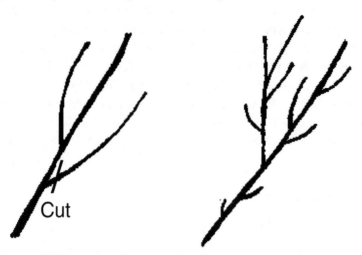

undisturbed. As a result, thinning-out cutting does not increase shoot growth to the same extent as heading-back.

c) Shortening cut

This type of cutting reduces branch length by cutting into older wood, usually at the point of a weak side shoot or a spur. Sea buckthorn sometimes will develop a heavy spur system with many weak and multi-branched spurs which have very limited shoot growth. Such trees tend to consistently produce poor quality fruits. A solution to this problem is to thin out some of the weaker spurs and head back some of the more extensively branched spurs (Fig. 7.4). This results in improved fruit size and quality.

d) Renewal cut

It completely removes older branches at their point of origin, usually leaving a short stub as a site for a replacement shoot.

Fig. 7.4. Responses of tree growth to spur cut. A. an unfruitful 'spur-bound' condition with numerous branches; B. Spur cut by removing branched spurs.

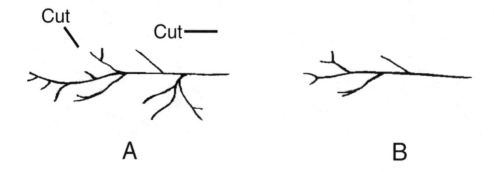

Fig. 7.5. Orientation of fruiting branches. A. Upright; B. Downward; C. Moderately vigorous and fruitful.

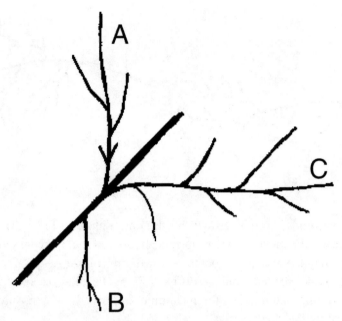

The pruning technique recommended for sea buckthorn is based on the idea that all branches are replaced in a specified order, usually over 3–4 years. How much should be pruned annually depends on the plant species, subspecies, or variety, the stage of growth, and planting spacing. In general, no more than a third of the top growth should be removed in one dormant season pruning.

In general, the upper and outer parts of the tree are more vigorous than the lower and interior parts. This is due to light exposure and orientation of the fruiting branches. The fruiting branches in any tree can be divided into 3 categories (Fig 7.5):

a) Upright growth is common in the light-saturated upper parts of the tree. It is excessively vigorous and fruit set is moderate and of poor quality.

b) Downward growth is common in the interior and lower parts of the tree. It often originates from the underside of larger branches. This type of growth is usually heavily shaded and sparingly set with poor quality fruit.

c) Horizontal branch growth, in most trees, this represents the bulk of the effective fruit bearing surface and this growth develops in areas of good light exposure. These branches are only moderately vigorous but very productive with superior quality fruit. Different areas within the tree have different pruning requirements and will respond differently. A pruning cut that is appropriate for a weak, heavily shaded branch may be inappropriate for vigorous, upright growth well exposed to light (Fig. 7.6).

There are 3 types of cuts to deal with undesired branches on the tree:

Fig. 7.6. Removal of excessive growth and shaded, downward branches. A. Unpruned; B. Pruned.

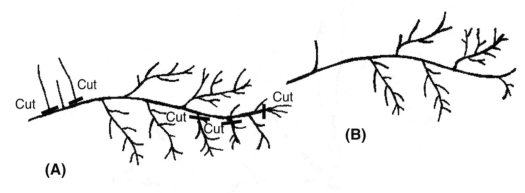

a) Remove vigorous, upright growth by thinning-out cuts (Fig. 7.3). To control or reduce vigor, remove the most vigorous, upright shoots and retain the less vigorous shoots (Fig. 7.5). These may have a greater potential for future fruiting.

b) Remove weak, shaded branches (Fig. 7.5) by heading-back cuts (Fig. 7.2) to enhance future fruiting potential. In the meantime, improve light exposure by removing the branches creating the shade.

c) Disturb horizontal branches as little as possible (Fig. 7.5). Most importantly, maintain good light exposure and avoid crowding, both within the branch itself and from adjacent branches with thinning-out cuts (Fig. 7.3).

Sea buckthorn may develop drooping, downward branches that cast interior shade and impinge on lower branches. In these cases, cut the drooping branch (Fig. 7.7). This drooping growth habit is often accompanied by a proliferation of fine wood near the

Fig. 7.7. Pruning drooping branches. A. Not pruned; B. Pruned.

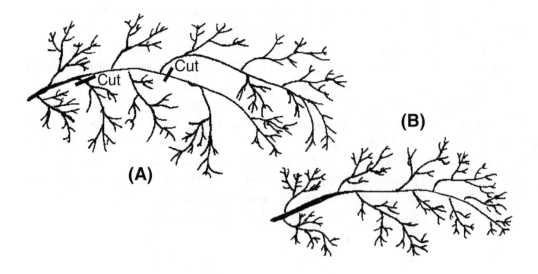

Fig. 7.8. Thinning-out fine wood caused by drooping growth. A. Unpruned; B. Pruned.

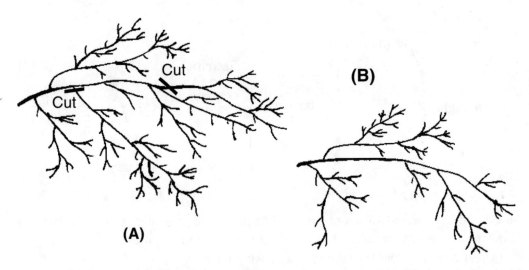

branch ends that requires regular application of thinning-cuts (Fig. 7.8). Crowding is one of the most persistent problems in sea buckthorn plantings especially in higher-density plantings. Maximum yields require a row space filled with effective leaf surface, but this will lead to some crowding. Overly aggressive pruning may stimulate excessively vigorous vegetative growth that creates more shade than before. In the attempt to control crowding, avoid extensive heading-back cuts. It is generally better to make relatively few but large cuts then to make many small cuts. Careful analysis will usually identify one branch that contributes most to the crowding in a given area. The removal of this branch alleviates crowding with minimal disturbance of the remaining bearing surface (Fig. 7.9).

Sea buckthorn produces fruit only on 2-year-old wood, usually in dense clusters held tightly around the fruiting stem. Since sea buckthorn is dioecious, male and female trees

Fig. 7.9. Elimination of overlapping branches on adjacent trees. A, Typical crowding within the row; B, Tree growth responses to heading-back cuts, which stimulate more growth; C, Eliminate crowding with thinning-cuts did not stimulate undesirable vegetative growth.

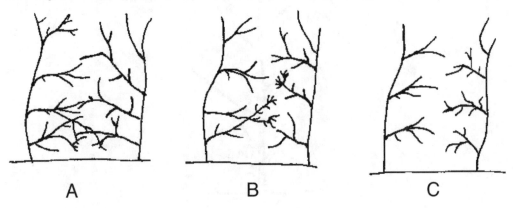

A   B   C

Fig. 7.10. Effect of terminal bud removal on tree growth. A. Terminal bud not removed; B. Same shoot as in 'A' at the end of the growing season; C. Terminal bud removed; D. Same shoot as in 'C' at the end of the growing season.

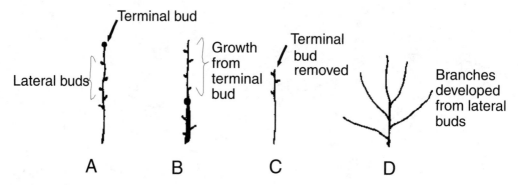

have to be interspersed within planting. Male plants are generally pruned differently to allow these trees to grow taller than females for pollination purposes.

Other considerations for pruning sea buckthorn are:

a) Remove the terminal bud at the beginning of the growing season or immediately after planting. As shown in Fig. 7.10, more branches are developed from lateral buds if the terminal bud is removed C and D). On the other hand, only single shoot will extend if the terminal bud is not removed (A and B),

b) Removal of superfluous growth on the lower part of the tree trunk will result in better development of framework branches above. As shown in Fig. 7.11, lower buds or weak branches should be removed.

c) Blind-buds may develop on branches without heading-back cuts. The results were significant between with and without heading-back cuts (Fig. 7.12, B and C).

d) Long and willowy branches have to be removed to stiffen the branches (Fig. 7.13).

Once sea buckthorn reaches the desired height, tree height must be maintained to ensure a balance of nutrients throughout the tree. Leaving trees too tall could result in lost

Fig. 7.11. Early removal of superfluous growth on the lower part of the tree.

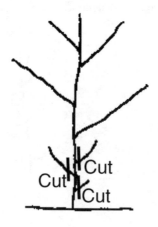

Fig. 7.12. Effect of heading-back cuts on the growth of branches. A. Branch with blind-buds; B. Branch with blind-buds, that was not headed back; C. Branch with blind-buds that was headed back.

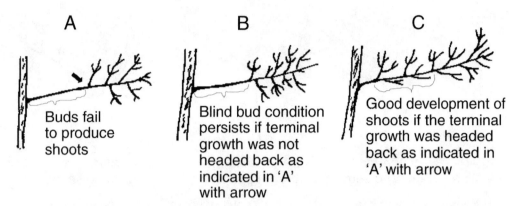

A

Buds fail to produce shoots

B

Blind bud condition persists if terminal growth was not headed back as indicated in 'A' with arrow

C

Good development of shoots if the terminal growth was headed back as indicated in 'A' with arrow

production in the lower part of the tree because of shading. Big branches are a problem in mature trees because of difficulties harvesting fruits, especially in the upper part. The branch: trunk diameter ratio may need to be reduced to 1:4 (25%), since the higher the planting density the lower should be the branch:trunk ratio. If vegetative growth gets out of hand, consider root pruning in March. Run a sharp blade 30 cm away from the trunk and 30 cm deep to cut feeder roots and curtail vegetative growth. The irrigation schedule may have to be modified if roots are pruned in this manner

There are two pruning systems to be considered in sea buckthorn orchards, the modified central leader and the open centre system.

Modified Central Leader

To encourage the development of a central leader, the vigor of the side shoots need to be reduced. After transplanting, trees are headed at approximately 60 cm, resulting in several upright, narrow-angled branches (Fig. 7.13) and the top shoot becomes the central leader during the first season. At the end of the second year, all new branches developing within 30 cm above the ground should be removed, and to maintain the dominance of the leader, all branches need to be headed back. Thinning cuts are also necessary. To restrain the tree at the desired height (not more than 2–2.5 m), the central leader must be cut back after 4 years of growth.

There are a few pointers worth considering for the modified central leader system:

a) "50% rule" for the central leader, leave no branches larger than 50% of leader diameter.

b) Remove excessive and old wood, enhance light penetration and renew wood even if less than 50% of trunk diameter.

c) "50% rule" on co-dominant branches, do not leave any side wood larger than 50% of the branch diameter.

Fig. 7.13. Open centre pruning system.

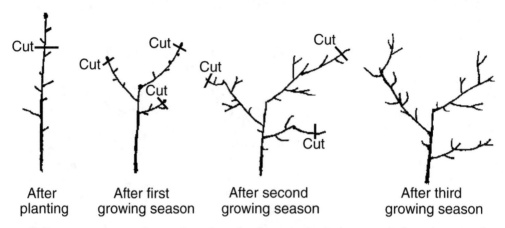

| After planting | After first growing season | After second growing season | After third growing season |

d) Remove spurs underneath and cut back extensively into weak, hanging wood.

e) Head back branches, stiffen up and keep tree compact with healthy wood close to the trunk.

f) Limit tree height only when tree is in full cropping state, ensure a low-vigor response at the top, but with lower part of tree open and vigorous enough to promote young, healthy flower buds.

Open Centre System

Normally, an open centre system has the characteristics of an upright growth, tree may have great vegetative extension and is usually not fruitful. However, in the case of sea buckthorn, with proper pruning to control its growth, this open centre system may be worth considering.

At planting, unbranched seedlings or rooted cuttings are pruned at approximately 60 cm and all buds within 30 cm from the ground were removed. Trees are trained to a multi-leader system with major branches extending up and away from the trunk, giving a vase-shaped tree. At the end of the second year, all newly developed branches within 30 cm above the ground have to be removed. Remove the potential central lead branch and head back and thin all branches to maintain at the desired height (< 2 m). Pruning has to be performed every year to avoid problems as the tree becomes older. The fruiting zone moves higher and to the outside of the canopy and the trees gradually assume an umbrella form.

Light penetration is very important to produce quality fruits. In this system, the tree canopy may produce shaded zones which eventually may cause poor fruit set. However, the important upper to middle section of the canopy should receive adequate sunlight and produce ample quantities of high quality fruit. Minimizing shaded zones should be considered during pruning.

Chapter 8. Diseases, Insects, Pest, and Weed Control

Thomas S.C. Li

Agriculture and Agri-Food Canada, Pacific Agri-Food Research Centre
Summerland, British Columbia, Canada V0H 1Z0

Sea buckthorn growers must protect trees from diseases and pests with carefully timed measures incorporating chemical or organic protection procedures. However, there has been very little research on disease and pest control on sea buckthorn in the world, and there are no chemicals registered for use in sea buckthorn orchards in Canada. Organic control measures are available, but regulations are different with individual organic growers associations, and growers should consult with their own association before application.

Diseases and insects are major factors affecting the success of sea buckthorn cultivation. For proper control of infestations, prevention and exclusion are the most important steps, with eradication as the last resort. Application of dormant oil before bud open is a very efficient control method. Exclusion measures, including quarantines and inspection of planting stocks, are designed to keep the pathogen from entering the growing area, or to reduce its presence to a minimum. Eradication measures are concerned with the elimination of the pathogen after it has become established in the area. All control methods are advanced by the eradication or removal of sources of primary inoculum, such as alternate or overwintering hosts, plant debris, or by field sterilization.

At the present time, sea buckthorn has relatively few diseases, insects and pests reported. However, it is likely more diseases and insects will be identified as sea buckthorn plantations grow in size and number. Verticillium wilt in sea buckthorn caused by *Verticillium albo-atrum* Reinke & Berth (Laurinen 1994) and *Verticillium dahliae* Kleb. (Kennedy 1987; Siimisker 1996; Piir 1996) have been widely recorded. *Verticillium dahliae* attacks a wide range of hosts including herbaceous and woody species. It was first reported in sea buckthorn in the U.S.S.R. (Petrova 1982). This disease is widespread among the areas where sea buckthorn is cultivated. It appears on fruiting trees at an age of 5–8 years. The symptoms of yellowing and progressive wilting of the leaves, the presence of discoloured vascular tissue also the non-symmetrical development of leaf symptoms are typical (Kennedy 1987). The infected fruits develop colour prematurely and dry up and shrivel. At the end of the summer, reddish swellings appear on the bark and truck cracks. An infected tree normally dies within 1–2 growing seasons. No effective control measures are known for verticillium wilt. Infected trees should be dug out and burned. Sea buckthorn should not be planted in same site for 3–5 years. Trees with signs of infection should not be used as a source of cuttings for propagation.

Fusarium wilt is another important disease in sea buckthorn. It is commonly believed that the fungus (*Fusarium* spp.) penetrates into seedlings through wounds on the roots or stem. A number of researchers believe that it begins with a drying out of a physiological nature, and that the fungus only attacks the rotting and dying plants. Infected branches are removed and burned as control measures.

Damping off of sea buckthorn seedlings occurs frequently when cotyledons or the first true leaves are open. It is believed that damping off can be caused by a number of soil-borne fungi, such as *Alternaria*, *Fusarium*, and *Botrytis* (Li, unpublished data).

Other diseases, such as die-back, are caused by *Fusarium* spp. (Kondrashov 1981*a,b*), sea buckthorn scab (Li, unpublished data), and a new pathogen found in China, *Phellinus hippophaeicola* Quél, was isolated from dead branches (Xu and Dai 1997).

Four recently noted insects infesting sea buckthorn platations in Canada are aphids, thrips, two-spotted mites and earwigs. The most damaging is the green aphid (*Capithophorus hippophae* Hille, Ris and Lambers) found in the new tender growth on shoot tips (Kadamshoev 1998). Green aphids stunt growth and cause the leaves to turn yellow, shrink along the centre vein, and finally to drop. Thrips are found in late spring and early summer. Earwigs are observed occasionally and two-spotted mites were found only once in the middle of a dry summer (Li, unpublished data).

Other insects damaging sea buckthorn are:

1) The gall tick (*Vasates* spp.), which damages leaves causing swellings or galls, as a result of which the leaf surface takes on a misshapen form.

2) The leaf roller (*Arhips rosana* L.) chews and rolls up leaves in May to July, and overwintering eggs were found on smooth parts of the bark of skeletal branches in late August.

3) The gypsy moth (*Ocneria dispar* L.) chews leaves deeply in the summer.

4) Commashaped scale (*Chionaspis salicis* L.) may cause heavy damage by sucking sap from the bark and cause the death of the plant.

5) Sea buckthorn fly (*Rhagoletis batava* Her.) is the most dangerous insect; it penetrates the fruits and feeds on the fruit flesh, making fruits unacceptable for use.

6) The caterpillars of the moth (*Gelechia hippophaeella* Schrk.) cause damage in buds by penetrating into and eating fresh buds.

A few insect infestations were reported in India. Death hawk moth (*Acherontia styx* Westwood) at the larval stage caused considerable damage in sea buckthorn. Some defoliating beetles, *Brahmina cariacea* (Hope) and *Brahmina* spp. 2, have also been found attacking sea buckthorn. Others such as *Holotrichia longipennis* (Br.) and *Plodia interpunctella* (Huber) caused damage at early growth and fruiting stages, respectively (Bhagat et al. 2001).

Damage from birds, such as ravens and magpies, is increasing and can result in the loss of entire crop. Deer, mice, pocket gophers, and other rodents occasionally can cause serious damage in sea buckthorn. The degree of damage depends on the population of the pests and the availability of other food sources. Deer damage occurs in both summer and winter from browsing on foliage and twigs, breaking limbs or trampling young seedlings. Deer will also eat fruit left on the shrubs during the winter months. Deer damage can be minimized by repellents or by providing alternate sources of food. Generally deer damage is not a serious concern on the Prairies; however, deer have caused damage, sometimes severely, in British Columbia (Li, personal observation).

Severe rodent damage usually occurs when trees are young. Rabbits girdle trees, debark limbs or chew new buds. They can be controlled by guards (wire, mesh, tree shelters), clean cultivation, or repellents. Field or meadow mouse damage generally occurs in winter when trees are girdled under the snowline. Damage is kept to a minimum if the tree row is weed free and grass between tree rows is regularly mowed. Livestock must be excluded from sea buckthorn orchards since livestock often destroy trees by browsing on foliage, breaking off limbs, trampling young seedlings or compacting the soil around the trees. Pocket gophers can severely damage trees by eating roots. Problems most often occur when orchards are bordered by grass or alfalfa fields. Pocket gophers can be controlled by trapping or poisoning.

Weed control is necessary especially for the survival and growth of newly planted sea buckthorn. Most weeds have a more vigorous root systems than trees and grow faster. Lack of adequate weed control causes more seedling mortality than any other single cause. Land preparation before planting is usually sufficient to eliminate weed seeds (see Chapter 4). Complete weed control is important until the sea buckthorn trees are large enough to shade out the weeds. This takes 4–5 years.

Mechanical or hand cultivation is very effective in controlling weeds adjacent to the tree row. Usually, three cultivations are required during the growing season. Cultivation should be shallow (8 cm) to avoid damaging the root system of the tree. Also, care must be taken to ensure the trees are not physically damaged during the cultivating process. In-row mechanical cultivation is accomplished by hand or using specialized in-row cultivating equipment.

There are several methods to control weeds. Deciding which one to use depends on the type of orchard operation (i.e., organic production), the weeds present, the soil type, and the equipment available. The choice of herbicide used in field shelterbelts depends on site conditions, tree and weed species, soil type, and climate. Seeds of annual weeds germinate primarily in the top 5 cm of the soil. Herbicides applied, either by pre-emergent, soil incorporated, and post-emergent, to the soil control weeds by inhibiting seed germination and development. Pre-emergent herbicides are applied to the soil surface, and rainfall or irrigation is necessary to move the herbicide into the soil for activation. Soil incorporated herbicides are incorporated manually after being applied to the soil surface. Post-emergent herbicides are applied to the foliage of weeds when they are small seedlings and growing actively. Such herbicides are absorbed primarily through the leaves and translocated throughout the plant. These herbicides are applied from late spring until early fall, depending on weed growth. Common symptoms on the weeds after treated with herbicides are curling or crinkling of new leaves with stunted twig development. Most serious herbicide damage occurs during spring spraying operations; it can be minimized by careful application in adjacent fields. Presently, there are no herbicides registered for use in sea buckthorn orchards.

There are no herbicides registered to control weeds in sea buckthorn orchards in Canada. However, herbicides; trifluralin, linuron, paraquat, glyphosate, sethozydim, simazine, and linuron and paraquat, are registered, if sea buckthorn is planted as part of shelterbelts (Anonymous 1994). In Germany, chlorpropham (5–7 kg/ha) was applied 4 days after planting with good results (Faber 1959).

Chapter 9. Sea Buckthorn Lipids

B. Dave Oomah

Agriculture and Agri-Food Canada, Pacific Agri-Food Research Centre
Summerland, British Columbia, Canada V0H 1Z

Composition of Sea Buckthorn Oil

Oil Content

Fruits

Sea buckthorn fruit consists of flesh (68%), seed (23%), and skin or peel (7.75%) (Zadernowski et al. 1997). In a more recent study of a large selection of sea buckthorn fruits collected from China, Finland, and Russia, Yang and Kallio (2001) found that the seeds constituted 23–28% of the fruits while the soft part (pulp/peel) represented the remaining 72–77% of the fruits. Three types of oils (seed, pulp, and peel oils) can be obtained from sea buckthorn fruit, since unlike other species, sea buckthorn synthesizes and accumulates fat in all morphological parts of the fruit. Sometimes the pulp and peel oils are not differentiated and hence referred to as fruit or pulp oil. The leaves also contain oil described later in this chapter. Oil content is a major index of the quality in sea buckthorn fruit. Thus, within the *H. rhamnoides* species, fruits of subsp. *rhamnoides* have significantly higher average oil content than those of subsp. *sinensis* (3.5 vs. 2.1%) (Yang and Kallio 2001). Oil content for most sea buckthorn varieties ranges from 1.7 to 6.6%; however, some cultivars grown in Russia have been reported to contain 9% oil (Schapiro 1989). Content of oil in the whole fruit varies widely, from 2 to 21% (w/w) on a dry basis depending on subspecies and origin (Zham'Yansan 1978; Yang and Kallio 2001) and is influenced by climate. Dry warm springs and autumns favor oil accumulation, while humid conditions, extended wet and cold weather, and shortened periods of sunshine leads to low oil content (Schapiro 1989).

Seed and Pulp

The oil content is generally high in both the seeds (up to 15% w/w, dry basis) and the pulp (soft parts; to 34% w/w, dry basis) of the fruit (Chen et al. 1990; Franke and Müller 1983*a, b;* Quirin and Gerard 1993; Wang 1990; Yang et al. 1992). Seed oil content ranges from 6 to 20% (w/w) (Beveridge et al. 1999; Yang and Kallio 2001) with seeds of subsp. *rhamnoides* containing a significantly higher oil content (11.3% w/w) than those of subsp. *sinensis* (7.3% w/w). Oil content in sea buckthorn seeds is influenced by paternal geno-

Fig. 9.1. Effect of altitude on oil content of sea buckthorn fruits (from Zhang, W. et al. 1989).

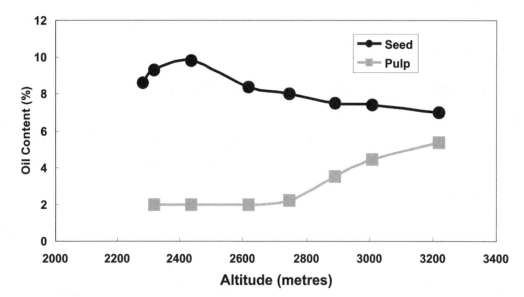

type, affecting the triacylglycerol by changing the expression of the C_{18} fatty acid, linolenic acid in the seed especially (Berezhnaya et al. 1989). Pollen origin also affects the acyl lipids in the mesocarp, or fruit soft tissue, thereby modulating the lipid content of the fruit. Growing altitude causes variation in both seed and pulp oils (Fig. 9.1) with pulp oil levels rising above about 2750 m (above sea level) and seed oil varying with altitude but oil content decreases above about 2500 m.

Differences in oil levels between pulp and seed oil decrease at high altitudes (Fig. 9.1, Yang 1989) and seed oil content also increases with latitude (Zhang, F. et al. 1989). The rate of oil accumulation in the pulp and seed varies depending on environmental (climatic and meteorological) factors. Pulp oil accumulates at a much slower rate than seed oil (Fig. 9.2) as fruit maturity proceeds, leading to different maximum time for accumulation of seed and pulp oils. Seed oil accumulates at a very rapid rate with the onset of maturation to a maximum and thereafter is constant or decreases as the fruit becomes mature and ripens. Pulp oil, on the other hand, rises more slowly over the maturation process and levels remain constant as the fruit reaches maturity, and ripens.

Major Components

Triacylglycerols (TAG)

Total triacylglycerol content (hexane extractives) of sea buckthorn fruit generally between 3–6%, varies due to location, and in some areas of the Pamirs in Russia can reach 8–14% (Mironov 1989). It comprises 85–90% of the oil, both seed and pulp oil, in the fruit (Yang and Kallio 2001). Ozerinina et al. (1987) using lipase hydrolysis to study the triacylglycerol composition and structure of whole fruit triacylglycerols found that the tri-

Fig. 9.2. Seasonal variations of oil contents of sea buckthorn fruits (from Yang et al. 1989).

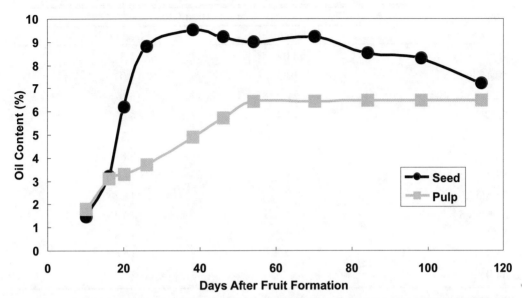

unsaturated (72–76 mol %) and the mono- and di-unsaturated (22–25 mol%) triacylglyc-erols predominated. Separation of seed and pulp oil triacylglycerols with capillary super-critical fluid chromatography (SFC) with a polar stationary phase, indicated that pulp oil consists mainly of 48-, 50-, and 52-acyl carbons about 69% of total, whereas seed oil con-sists of 54-acyl carbon triacylglycerol (about 74% of total) (Manninen et al. 1995). (Table 9.1).

Seed sea buckthorn triacyglycerol is similar to that of common oil seeds, such as flax and soybeans, in that it has a high content of oleic ($C_{18:1}$), linoleic ($C_{18:2}$) and linolenic ($C_{18:3}$) acids (Vereshchagin and Tsydendambaev 1995). These unsaturated fatty acids ($C_{18:1}$, $C_{18:2}$, and $C_{18:3}$ acids, 13–19, 47–50, and 32–37 mol %, respectively) display a concentration of fatty acids having a center of unsaturation at the sn-2 position (Ozerini-na et al. 1987). Since unsaturation the sn-2 position of a fatty acid is known to increase its *in vitro* absorption (Hoy and u 2001), sea buckthorn triacyglycerol has potential applica-tion in the delivery of energy and polyunsaturated fatty acids (PUFA) to persons suffering from oil malabsorption syndromes. In cases of normal absorption, sea buckthorn triacyl-glycerol may increase the rate of PUFA uptake for tissue regeneration or provide fatty acids for immuno-suppression using linolenic acid from sea buckthorn seed oil. This may explain the healing and recovery of burn patients by the topical application of sea buck-thorn oils which may be attributed to the increased demand for PUFA for tissue regener-ation.

Bound lipids, lipids not extractable by hexane or petroleum ether, also occur in sea buckthorn oils. The flesh contains the highest amount of total and bound (phospholipids and glycolipids) lipids, while the seed and skin account for considerably less (Table 9.2). Fruit skin and flesh carry a higher proportion of bound lipids to free lipids than seeds.

Table 9.1. Triacylglycerols of seabuckthorn pulp and seed oil (from Manninen et al.1995).

Compound	Proportion (%)	
major/minor(s)	Pulp oil	Seed oil
48:1	8.0	1.3
50:1/48:2	19.8	3.0
50:2/52:1, 48:3	21.6	3.6
52:2/50:3	19.2	4.6
52:3/54:2, 50:4, 56:1	9.8	5.2
52:4/54:3, 56:2	6.2	8.8
54:4/52:5, 56:3	3.9	11.7
54:5/52:6, 56:4	3.1	14.3
54:6/56:5	3.0	16.4
54:7	3.2	16.8
54:8	1.8	10.6
54:9	0.7	3.7

Table 9.2. Fat content (% dm) in different morphological parts of seabuckthorn fruit (from Zadernowski et al. 1997).

Fruit part	Fat content (% dm)		
	Free	Bound	Total
Whole	17.72	3.00	20.72
Flesh	16.10	8.40	24.5
Seed	11.32	0.91	12.23
Skin	9.86	3.77	13.63

Fatty Acid Composition

The fatty acid composition of sea buckthorn lipids has been investigated since 1929. The characteristic property of sea buckthorn fruit/pulp lipid is its high content of palmitoleic acid (16:1*n*-7) (between 20–45% of the total fatty acids) and the proportionality of palmitoleic acid to its precursor, palmitic acid (16:0) (Yang and Kallio 2001). Fruit/pulp lipid consists of 47% saturated fatty acids (mostly palmitic acid) and 53% unsaturated fatty acids (28% palmitoleic, 18% oleic, 4% linoleic and 2% linolenic acids). The fatty acid composition of triacylglycerol of fruit oil is dependant on the climatic and environmental conditions where seabuckthorn is grown. For example, the mesocarp triacylglycerol of sea buckthorn grown in the Caucases region differs from those grown in other regions in its higher level of $C_{18:1}$ accumulation (Vereshchagin and Tsydendambaev 1995). Fruit pulp/peel oil also contains a high level of palmitoleic acid (up to 55%), which is not common in the plant kingdom. This high concentration of palmitoleic acid may have cholesterol and triglyceride lowering as well as stroke-suppressing effects (Yang and Kallio 2000). Furthermore, improvement in the metabolism of vascular smooth muscle cells has been suggested due to an increase in the level of dietary palmitoleic acid intake (Yamori et al. 1986).

The proportion of palmitoleic acid in oil from the soft parts of sea buckthorn varies significantly among fruits of different origins (Chen et al. 1990; Berezhnaya et al. 1993; Ozerinina et al. 1987; Vereshchagin et al. 1998; Mamedov et al. 1981; Zhmyrko et al. 1984, 1987). Palmitoleic acid in the triacylglycerol of pulp/seed oil is reported to be higher in wild sea buckthorn fruit of Central Asia and the Baltics (55 and 42%, respectively) than in the Caucasian mountain regions (16%) (Vereschagin et al. 1998). Fruit pulp oil of wild subsp. *turkestanica* collected from two different locations in China also showed variation in palmitoleic acid (16 and 25%) (Chen et al. 1990). Differences in palmitoleic acid concentrations in pulp/peel oil also occur at the subspecies level with subsp. *sinensis*, *mongolica,* and *turkestanica* containing 31, 43, and 37%, respectively (Wang 1990). This palmitoleic acid variability in the oil from the pulp/peel (12–39%), and from whole fruits (9–31%) have recently been confirmed (Yang and Kallio 2001). Pulp oil, with this great variation in the proportion of palmitoleic acid, is attracting attention because of the increasing interest in the physiological role of monounsaturated fatty acids.

Palmitoleic acid is characteristic of the oils of the fruit coat and pulp, but is low in seed oils (Gao et al. 2000). Seed oil is characterized by high C_{18} unsaturated fatty acids (40% linoleic, 22% linolenic, and 17% oleic acids) and content only 21% saturated fatty acids (13% palmitic, 8% stearic acids) (Franke and Müller 1983). Seed oil is particularly rich in the two essential fatty acids linoleic (18:2n-6, up to 42%) and α-linolenic (18:3n-3, up to 39%) acids (Table 9.3). Seed oil from different fruit origins vary in linoleic (37–44%), α-linolenic (27–31%), palmitic (7–9%), stearic (2.5–3%), and vaccenic acids (2.2–2.8%) (Yang and Kallio 2001). The proportion of α-linolenic acid correlates negatively with those of oleic and linoleic acids in seed oil.

Table 9.3. Fatty acid composition (%) of oils from seeds, whole fruits and pulp/peel (from Yang and Kallio 2001).

Fatty acid	Seed oil		Fruit oil		Pulp oil	
	A	B	A	B	A	B
16:0	8.7[c]	7.4	22.9	23.6	26.7	27.8
16:1n–7	–	–	21.5	26.0[a]	27.2	32.8[b]
18:0	2.5	3.0[b]	1.5[b]	1.2	1.3[c]	0.8
18:1n–9	19.4	17.1	17.6	17.2	17.1	17.3
18:1n–7	2.2	2.8[b]	6.7	7.8[b]	8.1	9.1[b]
18:2n–6	40.9[b]	39.1	18.6	15.3	12.7	9.0
18:3n–3	26.6	30.6[b]	11.2[a]	8.8	7.1[c]	3.2

[a]0.05<0.1. [b]p<0.05. [c]p<0.01 between the two subspecies.
A – subsp. *sinensis*; B – subsp. *rhamnoides*.

Thus, sea buckthorn oil is a unique source of lipid for pharmacological and cosmetic use because of the high content of palmitic and palmitoleic acids in the pulp oils. The high content of polyunsaturates promotes its use in cosmetics, such as ointments, creams, and skin-care oils (Zadernowski et al. 1997), partly because of the ability of the polyunsaturated oils to promote absorption into skin as mentioned earlier (Hoy and u 2001).

Minor Components

Polar Lipids

The polar lipids of sea buckthorn fruit have received limited attention, even though they play important structural and physiological roles in cell membranes, and may offer interesting applications as emulsifiers and nutrients in cosmetic preparations. The flesh, seeds and skin of sea buckthorn fruits contain 3.3, 8.0, and 8.9% polar lipids, respectively. However, the types of fatty acids in the polar lipids from these different structural parts of the fruit are the same as the triacylglycerol fatty acids (Zadernowski et al. 1997). Recently, Pintea et al. (2001) investigating the fatty acid distribution in carotenolipoprotein complexes extracted from sea buckthorn fruits found that the polar lipids included phospholipids (61%) and galactolipids (39%). The fatty acid composition of polar lipids in the fruit is similar to that in the pulp and consist mainly of 18:1 (33.9%), 16:0 (31.9%), 16:1 (23%), and 18:2 (5.2%) fatty acids (Pintea et al. 2001).

Phospholipids

Sea buckthorn seed oil and whole fruit oil consist of 10–15% phospholipids (Kallio et al. 2000), while the fruit oil contains about 1% of phospholipids with lecithin as the major constituent. Eight phospholipid components have been detected in sea buckthorn seed oil (Goncharova and Glushenkova 1995*a*). Phosphatidylethanolamine (PE), digalactosyldiacylglycerol (DGDG), and monogalactosyldiacylglycerol (MGDG) together comprise 57% of the total polar lipid content of sea buckthorn fruit, while phosphatidylglycerol (PG) occurs only in small quantity (2.5%) (Pintea et al. 2001) (Table 9.4).

Palmitic, oleic, and linolenic acids are present in high concentrations reaching 26–27% of the total fatty acids, while myristic (14:0), palmitoleic, vaccenic (18:1n–11) and arachidic acids occur at very low concentrations. DGDG and MGDG are rich in oleic and linolenic acids and relatively poor in palmitic acids, while PA+DPG and PG are rich in 16:0 and poor in C_{18} acids (Pintea et al. 2001). Sea buckthorn meal is considered a high phosphatydylcholine (PC) lecithin since it contains up to 47% PC, much higher than the average soya lecithin at 24% PC or the high soya (PC) lecithin at 30–40% PC (Goncharova and Glushenkova 1995*a*).

Glycerophospholipids (GPL) fractionated from oil by silica Sep-Pak comprises 10–15% of total sea buckthorn fruit and seed oils (Kallio et al. 2000). However, GPL from whole fruit and seed exhibit large differences in fatty acid composition. Linoleic and α-

Table 9.4. Fatty acid composition of individual polar lipids of the carotenolipoprotein complexes extracted from sea buckthorn (from Pintea et al. 2001).

Fatty acid	Composition (%)						
	PC	PI+PS	PA+DPG	PG	PE	DG	MGDG
Total phospholipids							
& glycolipids (%)	9.6	10.5	19.7	2.5	18.6	15.7	23.3
14:0	3.9	5.2	5.9	3.8	8.3	2.9	3.3
16:0	17.4	20.3	26.8	22.8	24.7	10.2	12.2
16:1 (9c)	2.8	2.8	9.2	11.9	4.8	1.9	10.1
18:0	12.4	14.1	9.9	8.3	14.3	14.8	5.9
18:1 (9c)	17.9	19.7	19.9	21.5	15.3	26.1	11.3
18:1 (11c)	4.3	2.2	3.5	5.8	9.1	–	6.4
18:2	7.1	9.3	8.0	9.2	5.3	12.2	14.4
18:3	14.3	13.6	4.6	4.7	6.0	27.3	17.0
20:0	3.1	2.3	2.6	2.4	8.6	2.9	–
20:1	3.6	1.9	1.6	1.6	3.4	1.6	4.9
22:0	6.1	3.2	2.5	2.2	–	–	6.0
22:1	7.0	5.3	5.3	5.6	–	–	8.3

PC – Phosphatidylcholine; PI+PS – hosphatidylinositol+phosphatidylserine; PA+DPG – Phosphatidic acid+diphosphatidylglycerol; PG – Phosphatidylglycerol; PE – Phosphatidylethanolamine; DG – Digalacto-syldiacylglycerol; MGDG–Monogalactosydiacylglycerol.

linolenic acids together account for 60% of the total fatty acids in seed GPL, while in the whole fruits they constitute only 40% of the total fatty acids, with palmitoleic acid comprising more than 15% of the fruit oil GPL fatty acids. The linoleic and α-linolenic acid contents (46 and 15%, respectively) in the GPL of seed oil compare with the content of the triglacylglycerol fraction at 40 and 29%, respectively, of total fatty acids (Yang and Kallio 2001).

Tocopherols
The vitamin E content of sea buckthorn oil depends on its origin with those oils derived from the pulp after juice and seed removal (481 mg/100 g fruits) being higher than those from juice oil (216 mg/100 g of fruits) and seed oil (64–93 mg/100 g seed) (Beveridge et al. 1999). Alpha-tocopherol is the major tocopherol isomer in sea buckthorn fruit and seed oils representing 76 and 46%, respectively, of the total tocopherol (Table 9.5). Gamma-tocopherol is the second most predominant tocopherol isomer in seed oil constituting approximately 41% of the total tocopherol. The level of β-tocopherol is fairly constant at about 6.7% for both fruit and seed oils, while those of γ- (T_3) and δ- tocopherols are highly variable. The variability in tocopherol isomers encountered in seed oil of sea

Fig. 9.3. Synthesis of fatty acids during sea buckthorn fruit development (from Zadernowski et al. 1997).

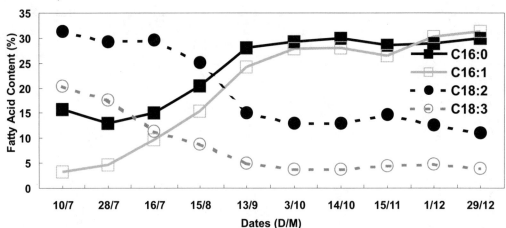

buckthorn cultivars suggest that selection for a particular isomer is possible. Extraction process also alters tocopherol profile (Table 9.5) indicating the possibility of using processing to obtain or modify sea buckthorn oil enriched in a particular tocopherol isomer. Oil extracted from seeds dried at low temperature soon after pressing of the fruit has been reported to contain vitamin E content as high as 3 g/kg (Lu and Ma 2001).

Carotenoids

Carotenoid content is one of the key characteristics by which sea buckthorn oil is traded commercially. Carotenoids vary widely depending on the source of the oil from 50 to 2139 mg/100 g (Beveridge et al. 1999). Pulp and fruit oils are the richest sources of carotenoids at about 5–10 g/kg of oil (Xin et al. 1995;Yang et al. 2001). Seed oil usually contains low levels of carotenoids at about 20–85 mg/100 g of oil. Mironov et al. (1989) indicated that sea buckthorn carotenoids consist of ~20% β-carotene, ~30% γ-carotene, ~30% lycopene, and ~15% oxygen-containing carotenoids. Up to 18 carotenoids have been identified in sea buckthorn fruit with the carotenoids possessing provitamin A activity (β-carotene, γ-carotene, β-zeacarotene, cryptoxanthin and sintexanthin) and lutein accounting for 48 and 14% of the total carotenoids, respectively (Kudritskaya et al. 1989). The presence of β-carotene, γ-carotene, δ-carotene, β- cryptoxanthin, and zeaxanthin in sea buckthorn fruit has been confirmed by thin layer chromatography, high performance liquid chromatography and mass spectrometry (Crapatureanu et al. 1996). Since carotenoids are powerful antioxidants and lycopene and lutein, in particular, are known to have physiologically beneficial effects in reducing the risk of certain cancers and in the prevention of age related macular degeneration, the potential of sea buckthorn oil as a nutraceutical and pharmaceutical ingredient is apparent.

Table 9.5. Tocopherol composition (%) of seabuckthorn oils (from Xin et al. 1995).

Tocopherol content	Fruit oil	Seed oil extracted by industrial methods			
		Pressed	Hexane	CO_2	FCHC
V_E (mg/100 g)	86.7	158.4	215.7	190.1	29.2
α-T	76.3	42.2	46.7	46.7	1.7
β-T	6.8	7.1	6.2	6.9	14.7
γ-T	7.4	43.1	39.7	41.5	58.9
γ-T3	6.3	1.4	2.2	–	–
δ-T	3.2	6.2	5.2	3.9	18.2
P-8	–	–	–	1.0	5.8

T– tocopherol; P-8 – plastchromanol -8; T_3 – tocotrienols; FCHC = fluorinated and chlorinated hydrocarbons; CO_2 = supercritical carbon dioxide extracted.

Sterols

The sterol content of fruit oil ranges from 2.2 to 8.8%, with fruit-coat lipids containing ~50%, pulp ~20% and seed ~30% of the sterols (Mironov et al. 1989). The structures of 17 sea buckthorn sterols have been established, β-sitosterol, β-amirin and α-amirin being quantatively, the most important (Mironov et al. 1989). The total sterol contents in oils from seeds, the fresh pulp/peel, and whole fruits are 12–23, 10–29, and 13–33 g/kg, respectively (Yang et al. 2001) (Table 9.6). Sitosterol constitutes 57–76 and 61–83%, respectively, of the seed and pulp/peel sterols. Sitosterol has recently been intensely investigated with respect to its physiologically beneficial effects in humans. Isofucosterol (also known as Δ-5-avenasterol) and obtusifoliol are the second most abundant phytosterols in the seed (about 15–17% of the total sterols), and together with stigmasta-8-en-3β-ol, constitute about 5–6 and 8–10% of the total sterols of pulp/peel and whole fruits, respectively. Earlier reports by Xin and coworkers (1995) indicated that the total sterol content of fruit is only one third to one half of that in seed oil.

In the pulp of sea buckthorn fruit, the main triterpene compounds consist of eight polycyclic alcohols: 24-methylenecycloartanol (13%), α-amyrin (2.7%), β-amyrin (3.2%), citrostadienol (0.4%), β-sitosterol (18%), 24-ethylcholest-7-en-3–ol (0.4%), erythrodiol (2.3%), and vivaol (2.3%) (Salenko et al. 1985). The triterpene fraction from the unsaponifiable part of sea buckthorn fruit oil consists of 24-methylcycloartene and 24-z-ethylidenecycloartanol (Glazunova et al. 1994).

Sea buckthorn fruit oil has 3–5% unsaponifiable matter with the fruit-coat, seed and pulp lipids containing 11, 1–2 and 0.3% unsaponifiables, respectively (Table 9.7). The unsaponifiables account for 3.5 and 4.1% of fruit and seed oil, respectively. The composition of unsaponifiables from oils of fruit is more complex than that from seed. Hydrocarbons in the unsaponifiable fraction of fruit-coat, seed, and pulp lipids amount to 50, 4, and 35% (wt basis), respectively, with 23:0 and 23:1 being the major hydrocarbons. Wax

Fig. 9.4. Extraction of sea buckthorn oil.

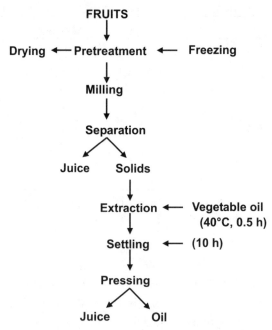

esters are found in the lipids of the fruit-coat (5–6%), seed (0.3%), and pulp (0.05%). The soluble volatile components of sea buckthorn oil consist of esters of aliphatic acids, aliphatic hydrocarbons originating from fruit wax, and bioactive constituents such as tocopherols and phytosterols which account for over half of the total volatiles (Chen 1989).

Biosynthesis

Fresh mature fruit contain 19% dry matter with lipid levels of about 2.8–7.8%. Sea buckthorn lipids are synthesized and accumulate gradually, the content increasing with fruit development (Fig. 9.2). The lipid content in fruit increases from 4.2% at the early stages of development (i.e., fruit setting) to approximately 17% (d.w) after 4 months. Polyunsaturated fatty acids (linoleic and linolenic acids), monounsaturated fatty acids (palmitic and palmitoleic acids), and saturated fatty acids are prominant at the early and late stages of fruit development, respectively (Fig. 9.3). The contents of C_{16} (16:0, 16:1) and C_{18} (18:2, 18:3) fatty acids increase and decrease, respectively, when the fruits change colour from pale green to yellow-orange. Thereafter changes in the concentration of these fatty acids are minimal until maturity (Zadernowski et al. 1997).

The reserve lipids of mature sea buckthorn fruits are characterized by an unusual fatty acid composition. The mesocarp of these fruits contains highly saturated oil consisting mainly of C_{16} acids, palmitic and palmitoleic acids, while the seeds accumulate triacylglyceride rich in unsaturated C_{18}, linoleic and linolenic acids. In the developing meso-

Table 9.6. Phytosterols in sea buckthorn (from Yang et al. 2001).

Components	Seeds	Pulp/peel	Whole fruits
Total sterols (mg/kg)	1200–1800	240–400	340–520
Total sterols in oil (g/kg)	12–23	10–29	13–33
Sitosterol (%)	57–76	61–83	60–62
α-amyrin and triterpenes (mg/kg)	22–56	58–317	57–436
Isofucosterol + obtusifoliol +	160–290	8–28	27–66
stigmast-8-en-3β-ol (mg/kg)	(15–17%)	(5–6%)	(8–10%)

carp, oil formation results in a constantly increasing oil level, whereas in the seeds after reaching a maximum, oil levels decrease sharply long before technical maturity of the fruits is achieved (Fig. 9.2). However, the relative rate of triacylglyceride formation in the mesocarp and the seed is similar at 1.28% per day, although the mesocarp exceeds the seeds considerably in duration of the triacylglyceride accumulation period. Accumulation of triacylglyceride in seeds generally proceeds more rapidly than in the mesocarp.

In green fruit triacylglyceride composition of mesocarp and seeds is similar indicating no differentiation in biosynthesis of fatty acids. During maturation, triacylglyceride accumulated by mesocarp and seeds are composed mainly of C_{16} and C_{18} fatty acids, respectively. The ratio of unsaturated to saturated fatty acids of triacylglyceride of mesocarp during the developmental stages remains unchanged while those in the seed increases drastically (Berezhnaya et al. 1993). In developing mesocarp, oil levels increase constantly, whereas in the seeds, after reaching a maximum the levels decrease considerably long before the fruit reaches maturity. During the development of mesocarp, the biosynthesis of palmitoleic acid ($C_{16:1}$) intensifies while the concentration of palmitic and other saturated fatty acids in the triacylglyceride decreases continuously. Two different patterns of triacylglycerol biosynthesis in the mesocarp oil of sea buckthorn has been observed (Verehschagin et al. 1998). These biosynthetic pathways lead to the different compositions of triacylglycerol of mesocarp oil related to the geographically diverse forms of sea buckthorn. Siberian, central Asian and Baltic forms containing mostly palmitoleic and palmitic acids in their triacylglycerol, are synthesized according to the 1,3-random and 2-random distribution patterns for fatty acids. The triacylglycerol of the geographical Caucasian form is rich in palmitic and oleic acids and conforms to the restricted random distribution of fatty acids.

The different mechanisms of triacylglycerol formation occur only in the mesocarp which is known to possess an exclusively maternal genotype. The distinction between the two mechanisms manifests itself primarily in the increased C_{18} and reduced C_{16} concentrations of fatty acids in the triacylglycerol of the geographical Caucasian sea buckthorn. This is probably due to the increased activity of 3-ketoacyl-acyl carrier protein synthetase (KAS II; EC 2.3.1.41). Decreased KAS II activity results in production of high concen-

Table 9.7. Composition of the unsaponifiable fractions of fruit-coat, pulp, and seed lipids of sea buckthorn (from Mironov et al. 1989).

Components		Fruit coat	Pulp	Seed
Unsaponifiables (wt%)		11	0.3	1-2
Hydrocarbon content				
(% of unsaponifiables)		50	35	4
Composition of hydrocarbons				
(% of unsaponifiables)	23:0	11	8	1
	23:1	15	9	–
	25:0	6	4	1
	25:1	6	3	–
	27:0	4	4	1
	29:0	7	7	1
Wax esters				
(% of unsaponifiables)		5-6	0.05	0.3
Composition of alcohol wax esters				
(%)	18:0	2	–	18
	20:0	9	7	1
	22:0	6	4	1
	24:0	8	4	1
	24:1	7	0.5	0.5
	26:0	7	–	1
	26:1	7	–	1

tration of C_{16} fatty acids in the mesocarp oils of sea buckthorn with Siberian, central Asian and Baltic origins. In the seed embryo only one system of triacylglycerol biosynthesis exists, regardless of the geographic origin of the sea buckthorn.

Extraction

Several methods of oil extraction from sea buckthorn have been described. Extraction of sea buckthorn oil usually involves pretreatment of the fruits (Fig. 9.4) to adjust moisture content followed by crushing, separation of the juice from suspended solids, extracting collected solids with vegetable oil with concomitant mixing and milling, and finally separation of the vegetable oil extract as final product. Juice clarification and addition of the collected juice solids to the press cake solids is recommended since it increases the carotenoid content, and hence the biological and commercial quality of the oil extract (Gavrishin et al. 1990; Bekaso 1992). Sea buckthorn oil can also be produced by drying the fruit to a moisture content of 1.2–1.5%, milling to a particle size of 500 μm, extraction with dichloro-difluoro methane at 0.6 MPa for 2–2.5 h and distilling the extract to obtain an oil with a carotenoid content of at least 500 mg % (Arkhipova et al. 1995). It is unlikely that dichloro-difluoro methane is acceptable as a food ingredient in many jurisdictions so local laws and regulations should be consulted before using this solvent.

Another method of producing a sea buckthorn oil is based on freezing the fruits at –22 to –27°C for 3–4 days, followed by thawing and fermentation over 2–3 days at 60–65°C and subsequent separation of juice. The flesh is then re-dried with continuous mixing prior to oil extraction. Separation of the end product from the dried mass is done by centrifugation (Nizhegorodtsvev and Umanskii 1997). In other methods (Horvarth et al. 1994), compounds promoting flocculation, such as egg white, are added to oil extracted from sea buckthorn flesh, the precipitate removed, and the oil dehydrated by using anhydrous sodium sulphate. The oil is then re-extracted by benzene (preferably hexane) and filtered for use for human consumption and cosmetic preparations. Since benzene is a know carcinogen, it is unlikely to be acceptable in any food related applications in most jurisdictions and so local laws and regulations should be consulted before following this methodology.

For simpler extraction, fruits are mixed with vegetable oil preheated to 50–60°C, milled with a centrifugal pin-type mill, and the seeds separated intact (Gorunzhina et al. 1990). During mixing of the milled seed with vegetable oil, it is treated with ultrasound (Kesariiski et al. 1990) for increased efficiency by increasing the ease of separation of the oil at later stages of processing. Sea buckthorn oil has been extracted with liquid (supercritical) carbon dioxide, both in the laboratory and at large scale, as early as 1978 by Shafton et al. (1978). In a later process, air dried pulp and seed (mass ratio 45:55) of fruits from plants grown wild in Siberia and Caucasus were ground (0.3–0.4 mm) and flaked to a thickness of 0.12–0.20 mm. Countercurrent extraction was then performed at 20°C and 57 MPa for 3–3.5 h to produce oil with a yield of 4–8 wt % depending on the composition of the feed material (Shafton et al. 1978).

In Germany, Flavex extracted sea buckthorn oil from Lithuanian derived plant material with supercritical carbon dioxide at 40°C and 35 Mpa. A 16.5 wt % yield of oil containing 29% palmitoleic acid (Flavex 1992) was obtained. The effect of extraction conditions on the rate of supercritical carbon dioxide extraction of oil from sea buckthorn seed and pulp has been investigated by Stastova et al. (1996). In this study, seed and pulp mechanically separated from air-dried press cake are ground and extracted in a semi-continuous extraction apparatus at 9.6–27 MPa and 25–60°C. The extracts obtained from seeds (8–13.4 mg oil/g seed) were different in colour (yellow or light orange) and consistency (viscous liquid) from that of the dark red pulp oil (yield 9.7–12.0 mg oil/g pulp) of very thick consistency. The extract consists mainly of triglycerides (76–92% of the total peak area) and only 1–4% of the free fatty acids. Carbon dioxide solubility of oil is dependant on variety and wheather seed or pulp derived. Pulp oil has higher solubility in carbon dioxide than seed oil probably due to its content of shorter fatty acids.

Table 9.8. Glycolipids of seabuckthorn leaves and their fatty acid compositions (from Goncharo-va and Glushenkova 1995*a,b*).

Fatty acid	Composition (%)					
	SQVDGs	DGDGs	X$_1$	X$_2$	MGDGs	AsSGs
Amount,						
% by weight	7.5	32.5	1.2	1.7	45.2	6.2
14:0	3.3	0.3	0.4	–	0.3	1.4
15:0	1.6	0.2	–	0.4	0.2	0.8
16:0	55.1	21.4	18.5	20.0	0.1	28.3
16.1 (3) + 16:1 (9)	3.2	1.5	1.5	1.8	1.3	2.9
16:2	–	1.3	–	–	–	–
16:3	4.8	1.1	1.1	1.1	0.9	0.9
18:0	6.1	1.2	–	–	–	1.4
18:1	6.0	1.4	6.1	8.3	2.9	14.4
18.2	3.6	2.7	10.4	9.0	11.7	10.8
18.3	15.8	64.2	58.0	57.7	72.2	31.4
Other acids	–	1.7	4.0	1.7	–	7.5
Σ sat	66.1	25.1	20.1	20.9	–	33.7
Σ un sat	33.4	73.9	79.9	79.1	–	66.3

Σ sat, Σunsat - Sum of saturated and unsaturated fatty acids, respectively. SQVDG = sulfoquinovosyl-diglycerides; DGDG = digalactosly-diglycerides; MDGD = monogalactosyl-diglycerides; AsSG = esteri-fied sterol glycosides; X$_1$, X$_2$ = unidentified glycolipids.

Leaf Lipids

Air-dried leaves of sea buckthorn contain 10–12% chloroform:methanol (2:1) extractable lipids, 50–57% of which is reextractable by hexane (Goncharova and Glushenkova 1996). Leaf lipids consist mainly of waxy compounds with wax esters, iso-prenols, and sterols contributing 33, 35, and 8.8%, respectively of the total lipid. The hydrocarbons containing C_{29}, C_{30}, and C_{31} represent approximately 57.5, 3.7, and 20.5%, respectively, of the total hydrocarbon and together account for 81% of the total hydrocarbon composition of sea buckthorn leaves.

Sea buckthorn leaves contain almost equal amounts of neutral and polar lipids (51.4 and 48.6% of the total extract). The polar lipids consists of glycolipids (36.5%,w/w) and phospholipids (12.1%, w/w) of a total chloroform:methanol extract (Goncharova and Glushenkova 1995*a,b*). Monogalactosyl-, digalactosyl-, and sulfoquinovosyl-diglycerides (MGDG, DGDG, and SQVDG) are the main classes of glycolipids (GL) together amount-ing to 85% of the weight of the GL (Table 9.8). These three classes of glycolipid, MGDG, DGDG, and SQVDG occur in the 6:4:1 ratio (by wt) similar to those found in leaves of other higher plants. MGDG and DGDG are rich in linolenic acid (72 and 64% of the total

Table 9.9. Phosopholipids of seabuckthorn leaves and their fatty acid compositions (from Goncharova and Glushenkova 1995*a,b*).

Fatty acid	Composition (%)			
	PC	PE	PG	PI
Amount, %by weight	32.8	29.7	21.5	16.0
14:0	0.6	1.2	0.8	1.0
16:0	30.4	35.2	21.5	37.4
16:1	0.8	2.5	12.6	1.9
18.0	2.5	2.2	1.2	3.6
18.1	5.3	4.8	2.4	4.2
18.2	19.5	15.2	8.5	16.8
18:3	38.2	32.3	48.6	27.5
Other acids	2.7	6.6	4.4	7.6
Σ sat	34.7	40.0	25.5	44.6
Σ un sat	65.3	60.0	74.5	55.4

Σ sat, Σ unsat - Sum of saturated and unsaturated fatty acids, respectively.
PC = phosphatidylcholine; PE = phosphatidylethanolamine; PI = phosphatidylinositols; PG = phosphatidyl-glycerol.

fatty acids, respectively) while palmitic acid (over 50% of total fatty acid) predominates in SQVDG. Phosphatidylcholines (PC), phosphatidylethanolamines (PE), phospalidylglycol (PG), and phosphatidylinositols (PI) are the main classes of phospholipids identified in sea buckthorn leaves. These phospholipids have similar fatty acid compositions (Table 9.9) with palmitic, linoleic and linolenic acids being most predominant. The unsaponifiable fraction of an ethanol extract of sea buckthorn leaves has been investigated in an attempt to developing an analog of sea buckthorn oil from cheap raw material (Gonchareva and Glushenkova 1995*a,b*). It contains the same group of compounds as the unsaponifiables of the pericarp lipids together with phytol, campesterol, cycloartenol, nor-triterpene alcohols and polyphenols. Epicuticular waxes and cell lipids account for 0.47 and 3.53% of the weight of the initial leaves, respectively. The polyphenols, known to have biological activity, amount to 12% of the weight of the unsaponifiable substances of the leaves. A freon extract (2%) of the leaves in sunflower oil exhibits stimulating effect on skin regeneration process.The epicuticular lipids consist mainly of esters of long-chain alcohols and higher fatty acids that together make up more than half of the weight of the whole extract. Behenic, $C_{22:0}$, arachidic, $C_{20:0}$, lignoceric, $C_{24:0}$ and palmitic acids are the most abundant fatty acids and tetracosanol, $C_{24:0}$, and docosanol, $C_{22:0}$ are the main alcohols of the alcohol fraction. The alcoholic fraction also contains 12% cyclic alcohols, consisting mainly of β-sitosterol, epifriedelanol and 24-methylenecycloartanol. Triterpene acids and hydrocarbons account for 16 and 10%, respectively, of the epicuticular lipids.

The triterpene acids consist of ursolic and oleanolic acids, representing 90 and 10%, respectively, of the weight.

The epicuticular lipids like the cell lipids are represented mainly by the class of esters, but in the lipids of the surface wax of the leaves more than half the weight of the total lipids (58.7%) consists of esters of long-chain alcohols and acids, while in the cell lipids (44% of the weight of the total extract) consists of esters of fatty acids and various alcohols mainly cyclic and isoprenoid. The main saturated fatty acids of epicuticular and cell lipids are C_{20-24} and $C_{16:0}$, $C_{16:1}$, respectively.

The compositions of sterols and triterpenes from sea buckthorn leaves have been studied by Salenko et al. (1985). The neutral lipid contains a fraction (23.4% of the neutral lipids) consisting of mixtures of hydrocarbons, carotenoids, tocopherols, polyphenols, phytol, aliphatic and polycyclic alcohols (sterols and triterpenoids). The aliphatic alcohols (19%) are represented by *n*-alkanols, predominantly C_{24}(8.6%), C_{22}(4%), C_{26}(3.5%), C_{20}(1.4%), C_{28}(0.8%), C_{18}(0.4%), C_{25}(0.2%), and C_{23}(0.1%). The polycyclic alcohols consist mainly of a group of triterpene compounds: dimethylsterols (11.6%), methyl sterols (0.8%), sterols (6.8%), triterpene aldehydes (1.1%), and diols (9.1%). Five components: α-amyrin, β-amyrin, 24-methylenecycloartanol, cycloartanol, and lupeol are present in the monohydric triterpene alcohols and dimethylsterols. The methylsterol fraction consists of obtusifoliol, citrostadienol, β-sitosterol, erythrodiol and uvaol. The aliphatic alcohols of sea buckthorn leaves has been reported to consist of 14 triterpenoid compounds identified as cycloartenol (2.6%), 24-methylenecycloartenol (1.3%), α-amyrin (3.4%), β-amyrin (1.7%), lupeol (2.6%), obtusifoliol (0.3%), oleanolic aldehyde (0.5%), ursolic aldehyde (0.6%), β-sitosterol (6.8%), nordiene I and II (0.2 and 0.3%, respectively), erythrodiol (5.8%), uvaol (3.3%), and citrostadienol (0.5%).

Current Trends

The contribution of a large number of physiologically active substances by sea buckthorn oil has prompted research in several areas of potential application as claimed by numerous recent patents (Table 9.10). These patents can be classified based on target areas into processing (three patents), cosmetic (seven patents), and pharmacology (four patents). According to these patents, the beneficial protective effect of sea buckthorn oil toward inflammation and skin damage, together with its anti-allergic, bactericidal, antiseptic and regenerative properties make it an ideal product for cosmetic applications. Improvement in the composition of sea buckthorn oil by mixing with juice (Li et al. 1996) or by refinement (Xing and Wang 2000) enhances its quality for very high end skin care, anti-aging, and anti-wrinkle products. The wound-healing and antioxidant properties (Libman and Zolotarsky 2001), blood lipid management effects (Zhou 1999), and endocrine regulating effects (Wei and Cao 1995) broaden the use of sea buckthorn oil in the area of prophylactics and pharmacology. Hence, processes have been designed to obtain oil pure

Table 9.10. Review of Sea Buckthorn Patents.

Claims	Patent Reference
Unsaturated fatty acids of sea buckthorn seed oil regulates blood lipids, resist angiocslerosis and radiation, restrains tumour cell, strengthen immunity, and nourishes skin.	CN1207920 Zou (1999)
Sea buckthorn fruits processed by comminuting the fruits, fermenting the pulp with enzymes to release pulp oil and separating the oil by centrifugation. The oil is claimed to be useful in cosmetic, pharmaceutical, and food products.	DE4431393 Lorber and Heilscher (1996)
Oil extract of sea buckthorn exhibits regenerating, softening, wound-healing and protective effects against external actions and insect stings (for skin care products).	RU2106859 Senjavina et al. (1998)
Sea buckthorn oil in cosmetic cream increases biological activity and broadens variety of cosmetic creams.	RU2134570 Bencharov (1999)
Ointment containing sea buckthorn (0.5–1.5%) suppresses caragenin - induced edemas and passive cutaneous anaphylaxis (treats patients with inflammatory and allergic skin damages).	RU2132183 Prokof et al. (1999)
Ointment containing sea buckthorn oil is antiseptic and is intended for treatment of burns and infected injuries.	RU2129423 Frolov (1999)
Cosmetic cream containing sea buckthorn oil protects face skin in winter.	RU2120272 Detsina and Selivanov (1998)
Cream containing sea buckthorn oil is antiallergic, bactericidal, anti inflammatory, regenerative, and biologically active.	RU2123320 Chistjakov (1998)
Process for producing sea buckthorn seed oil by heat pretreatment (10-30 min. microwave), or infra-red drying prior to processing, settling, and filtering.	CN1123318 Li et al. (1996)
Capsule containing sea buckthorn fruit oil regulates endocrine function of women by raising female hormone levels, curing climacteric syndrome, early aging of ovaries, and premenstrual syndrome.	CN1106691 Wei and Cao (1995)
Liquor containing sea buckthorn juice and oil has high potassium level and anti-aging properties.	CN1104054 Li et al. (1996)
Processing method to obtain sea buckthorn seed oil with high oil yield and low residual solvent.	CN1089301 Xin and Zhou (1994)
Sea buckthorn essence made up by mixing refined sea buckthorn oil increases immunity, promotes microcirculation of the skin, and delays skin senility.	CN1260170 Xing and Wang (2000)
Composition including sea buckthorn oil extract and antioxidant maintains its distinctive colour and may be used as a cosmetic, pharmaceutical, nutraceutical, or a food product.	US6187297 Libman and Zolotarsky (2001)

enough to meet the standards of these industries, and be economical, efficient and effective for use for human consumption.

China's Ministry of Health has already designated sea buckthorn seed oil as a health (functional) food, and the Chinese Green Food Development Centre lists it as a grade AA green food. In the U.S.A., sea buckthorn oil is becoming popular as a nutritional supplement to treat the gastrointestinal disorders that afflict more that 70 million people in the U.S.A. according to the Centres for Disease Control (Papanikolaw 1999). For this purpose, a Pennsylvania-based company, Flocare Medical, is said to market Ulcer-EZZZ™ soft gel capsules containing 500 mg of sea buckthorn seed oil extracted using supercritical carbon dioxide. Clinical studies performed by Flocare Medical show that sea buckthorn seed oil inhibits gastric ulcer formation at a higher rate (65.5% inhibition) compared to regular treatment with the drug cimetidine at 200 mg (55.2% inhibition), and control (0% inhibition) (Business Wire 1999).

Products containing sea buckthorn oil are already being widely marketed and distributed by the cosmetic industry. Draco Natural Products has targeted the personal-care market, including applications in lotions, creams, hair care products and sun-care protection products for the launch of its sea buckthorn products (Anonymous 2001a). A&E Connock in Hampshire, U.K. markets sea buckthorn oil as a cosmetic and perfumery base, while Rahn Cosmetics, Middlesex, U.K., supplies sea buckthorn gels as a cosmetic ingredient. Tagra Biotechnologies Ltd. has commercialized sea buckthorn oil using its proprietary microencapsulation technology for its line of vitamins and natural products (Anonymous 2001b) . New products containing sea buckthorn oil will continue to emerge as knowledge and understanding of this interesting oil becomes clearer and increased focus is brought to bear through well designed clinical trials to support claims.

Chapter 10. Post Harvest Handling and Storage

Thomas H.J. Beveridge

Agriculture and Agri-Food Canada, Pacific Agri-Food Research Centre
Summerland, British Columbia, Canada V0H 1Z0

The beginning of the transformation of the raw sea buckthorn fruit to a sophistocated product requires appropriate mechanical harvest, transportation, holding and storage procedures. Fruits are hard to harvest and persist on the branches all winter due to the absence of an abscission layer. Furthermore, most, but not all, varieties of sea buckthorn carry long, spiny thorns, which make manual harvesting slow and unpleasant, and this combined with the small fruit size makes mechanical harvesting mandatory for commercial purposes, other than for the existing small fresh market. This is particularly true in a North American setting, where labor costs tend to be high. In Saskatchewan, the total labor cost for harvesting an orchard of 4 ha was estimated to be 58% of the total cumulative production costs over 10 years.

In spite of these costs, manual methods are still practiced, sometimes augmented by a scissors-like comb device operated by the picking hand to comb the fruits from the branches into a receiving vessel. In Asia, harvesting is done by hand, and this difficult and labor intensive process requires about 1500 person-hours/ha (Gaetke and Triquart 1992). Wolf and Wegert (1993) and Orlander (1995) reported that difficulties associated with harvesting are major barriers to orchard production in Europe. Therefore, development of sea buckthorn as a cash crop depends upon the development of mechanical harvesting methods. Koch (1981) harvested entire fruiting shoots with pneumatic shears, and Botenkov and Kuchukov (1984) developed a device for hand picking that consists of two hinged jaws with teeth and brushes. Savkin and Mukhamadiev (1983) designed a pruning machine to trim sea buckthorn into a hedge more suitable for mechanical harvesting. Other mechanical harvesters have been developed in Sweden, Germany and Russia (Orlander 1995), but most have disadvantages, such as fruit and bark damage and low efficiency. Harvesting methods tested include shaking (Affeldt et al. 1988; Gaetke et al. 1991), vacuuming (rapid air flow) (Varlamov and Gabuniya 1990) and quick freezing prior to fruit removal (Gaetke and Triquart 1992).

Many harvesters are based upon the principle of cutting off the fruit branch (Gaetke and Triquart 1992; Olander 1995). Olander (1995) developed an over-the-row mechanical harvester which removes the fruit-laden branches and extracts the fruit by shaking the branches in an axial direction. Considerable German work on mechanical harvesting of sea buckthorn has involved removal of the fruiting branch, freezing the fruits (–36°C overnight under natural winter conditions in the tests described), then beating the branch-

Fig. 10.1. Percentage of fruits removed during two 15 second shaking episodes for harvest trials conducted in Saskatchewan, November 2–3, 1999 (from Mann et al. 2000, with permission).

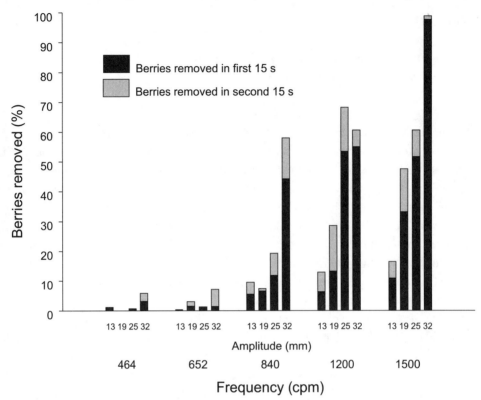

es to remove the fruit (Wolf and Wegert 1993). Gaetke and Triquart (1993) developed a machine in which fruiting branches are hand-fed into the harvester, and the fruit is separated from branches and leaves by screen conveyers and fans. A serious disadvantage of removal of the fruiting branch is that since sea buckthorn sets fruit on second year wood, harvesting by cutting the fruiting branches means that a harvest can be obtained only every 2 years.

The most successful methods of mechanical harvesting have involved shaking the individual branches of the shrub *in situ* to dislodge the fruits into a catcher placed around the base of the tree. From here the fruits are placed in bins or baskets for transport to the storage or processing facility. In Saskatchewan, Mann et al. (2001) developed a tractor mounted and powered branch shaker and tested the frequency and amplitude requirements for harvesting sea buckthorn at various harvest dates using a specifically designed shaking device that operated on branches removed from the tree. Shaking frequencies less than 1000 cycles per minute (cpm) were ineffective regardless of amplitude. In November, following 3 days with average minimum temperature of –7.7°C, 99% of the fruits were removed using a shaking frequency of 1500 cpm and a shaking amplitude of 32 mm (Fig. 10.1). Harvesting prior to this date was less effective and during January testing (winter

Fig. 10.2. Sea buckthorn fruit picked without pedicel, and frozen. Torn tissue due to pedicel removal can be seen around the top circumference. (from Harrison and Beveridge 2002, with permission).

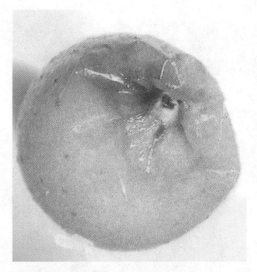

in Saskatchewan) the branches were brittle and many broke when shaken at 1500 cpm. Usually shaking beyond the initial 15 seconds did not cause many additional fruits to be removed. These tests suggest that harvesting after the first winter freeze or snow would be a suitable time for harvest. Considering these results in conjunction with the German reports already mentioned suggest that freezing may be the important trigger for facile fruit detachment. However, this late harvest will affect fruit quality.

Removal of fruits by shaking requires the facile separation of the fruit from the branch. In sea buckthorn this requirement is not well met and the fruits can cling tenaciously, resisting vibratory removal. Microscopic examination of the structure of the fruit and its attachments (Harrison and Beveridge 2002) indicated that the skin (epidermis) of the fruit is confluent with the pedicel or stem making fruit removal problematic. Commonly, manual or machine removal of fruits damages the pedicel (stem) end of the fruit (Fig. 10.2), providing for leakage of juice and entrance of wash water as will be discussed later. Insertion of an natural break or abscision layer through the stem of the fruit would promote more facile fruit removal, but such a structure does not exist in sea buckthorn. However, careful study has revealed the existence of a natural, apparent abscision zone in the interface between the pedicel and raceme structure through which the fruit is attached to the stem (Fig. 10.3). The existence of this zone may explain the observation that mechanical harvest at the right time or maturity often provides relatively easy harvest of fruits with stem attached and the variation between trees in susceptibility to mechanical harvest. Also it provides a target for plant breeders to improve sea buckthorn varieties by enhancing this zone over a broader range of maturities.

Mechanical harvesters have also been designed in handheld versions. They are vibrational in operation and still need improvement (Fig. 10.4). One serious disadvantage of the

Fig. 10.3. Sliced, frozen, ripe fruit and raceme showing the arc of separation (apparent abscision zone) at the intersection of the pedicel and raceme. (p–r). Labels as r = raceme, p–r = pedicel raceme interface, p = pedicel (stem), p–f = pedicel-fruit interface, f = fruit (from Harrison and Beveridge 2002, with permission).

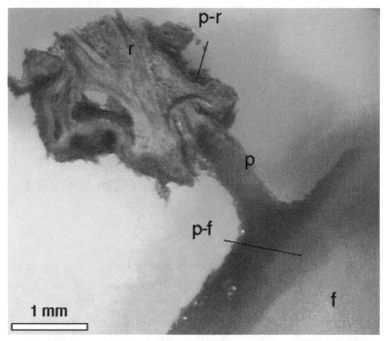

handheld machines is the transfer of the vibrations to the operator and the resulting rapid development of operator fatigue. These machines have been tested in British Columbia with some success, but the ideal amplitude and frequency is not known with accuracy. Instead the machines are adjustable and the parameters are adjusted to suit the conditions, but the results reported from Saskatchewan will help in making these adjustments. Some skill and care is required in the operation of these hand-held harvesters (and probably also the tractor-mounted versions) to avoid bark damage (Fig. 10.5). Clearly, tractor mounting the vibrational device would eliminate the operator fatigue factors, but the machine becomes more clumsy in use. Harvesting in British Columbia is different from harvesting in Saskatchewan, because there are only occasional winters where the temperature conditions of the prairies are obtained, although temperatures of –10°C in early winter occur frequently and may suffice. Harvesting methods other than waiting for winter freezing will be necessary for regions with moderate climates, and the moisture losses involved with waiting for the first seasonal freeze may be unacceptable in some circumstances. It will be necessary to determine optimum harvesting indices for the various growing areas, and for particular varieties of fruit.

Hormonal treatments to reduce the force required to detach the fruits have been tried and look promising (Li and Schroeder 1996). Trushechkin et al. (1973) reported that Ethrel (ethephon) at 2000 mg/L of water decreased fruit detachment force by 30%. Ethep-

Fig. 10.4. Hand held oscillatory harvester which has proven useful as a plot harvester in the Okanagan Valley of British Columbia. The machine is still under development to reduce operator fatigue from vibration.

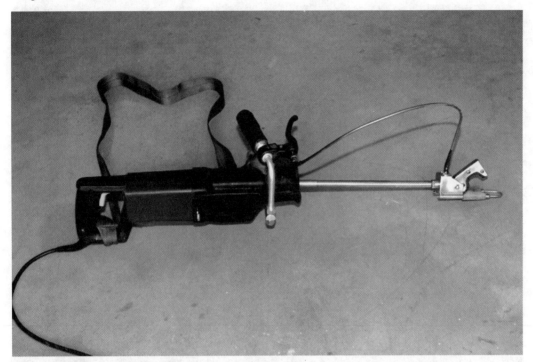

hon is normally applied 7 days before harvest, and each harvest date must be determined locally depending upon local conditions and orchard practice. Demenko et al. (1986) suggested that the inability of ripe fruit to abscise is caused by compartmentalision of internal ethylene in seeds. Ethylene treatment induced the formation of an abscission layer (Demenko et al. 1986; Demenko and Korzinnikov 1990). Reduction of the force required to remove fruits might provide impetus to the use of mechanical harvesters adapted from other crops, such as blueberries or raspberries. At present these machines do not work. The further effect of hormones, such as ethylene, on the harvesting, storage and processing character of sea buckthorn fruits have not been determined, but fruit ripening correlates with higher internal ethylene concentrations in the seed (Demenko et al. 1986). This would suggest that the fruits might display a climacteric respiration pattern which would have significant consequences for the subsequent storage and processing of the fruit. However, Zhang, W. et al. (1989*b*) determined respiration patterns in sea buckthorn fruit ripening that were non-climacteric.

The fruits will remain on the branches all winter, if left undisturbed (Li and Schroeder 1996), and this might be a good way to provide for short-term storage. Some harvesting schemes make use of this factor combined with natural freezing processes as discussed above. However, dehydration of the fruits will cause loss of juice product if the time allowed is excessive, and late harvest will lower vitamin C and total carotenoid levels

Fig. 10.5. Bark damage on a sea buckthorn branch caused by improper holding of the branch during shaking. It is important that the branch be held firmly in the machine "jaws" to avoid this damage caused by impact of the "jaws" on the branch.

while reducing soluble solids (Li, unpublished data). On the other hand, Tang (2002) indicated that the increase in fruit fresh weight correlated strongly and significantly with decline in concentrations of vitamin C and titratable acidity.

Harvesting the fruits, by either mechanical or manual means, is problematic in that the fruit does not release easily from the stem in most varieties. This means that the skin is often torn when the fruit is removed by violent shaking by mechanical harvesters or by pulling and squeezing by pickers. If a fruit is removed from a branch in such a way as to leave the stem on the branch, or if the stem is removed from an otherwise intact fruit, gentle pressure will cause oozing of a clear or cloudy fluid from the stem end of the fruit. Sometimes the gentle pressure will cause bursting of the fruit where the stem originated and release of a cloudy or slightly yellow fluid. This ruptured skin provides entry for spoilage organisms and insects during subsequent transportation and storage. It also provides an exit for fluid to ooze from the fruit during subsequent handling, storage and

Fig. 10.6. Juice which has exuded from skin-damaged fruits during several hours air transportation from Saskatchewan to British Columbia and storage at 1°C on arrival. Fruits were hand picked, packed in plastic bags, and placed in bins to a depth of about 1 foot for transport. The fruit exudate was a clear, essentially transparent fluid containing a sugar profile very similar to that of pressed juice and with an oily surface resulting from pulp oil exuding with the fluid.

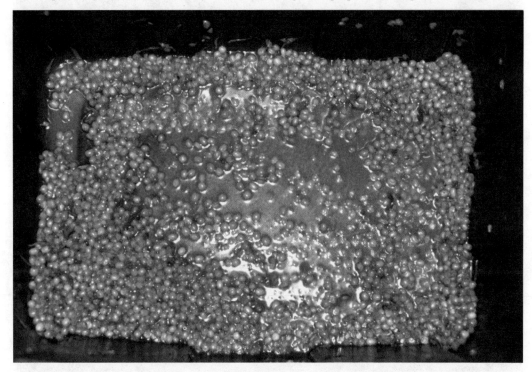

transportation operations (Fig. 10.6). This fluid loss may result in lower juice yields from the fruits and can give the bin of fruits an unsightly appearance while providing a wet surface for the adhesion of dust, dirt and debris.

Both hand picked and mechanically harvested fruits (Fig. 10.7) carry harvesting debris so that some degree of inspection and cleaning is required. The purpose is to remove diseased or pest-infested fruits and stems, leaves and other debris collected during harvesting. This debris would be expected to contribute off-flavors from rotted fruits or "green" stem or leaf flavors from vegetable residues. Prior to further processing, the harvested fruits must be passed over a well lit inspection table for removal of visible waste material. Cleaning to this level is strongly advised. Washing the fruit prior to juice extraction is important for removal of microorganisms and dust and dirt clinging to the fruit (Lui and Lui 1989). This is a common procedure when processing fruit, such as apples, which have a tough water impermeable skin (Bump 1989), floated in stirred tanks where a detergent or wetting agent is employed to assist the cleaning process.

Sea buckthorn fruits are well known to carry a "musky" odor, detectable even in the sea buckthorn field and washing may reduce this odor (Beveridge et al. 1999). Sea buckthorn fruits have been washed in air-agitated tanks (Heilscher and Lorber 1996b), and

Fig. 10.7. Shaker-harvested sea buckthorn fruits showing contamination with harvesting debris. Most of the debris is leaves, but some broken branches are present. Removal of these leaves would be desirable, as they may provide some carry-over flavor and they tend to clog pumps.

40°C is suggested as a washing temperature (Zhang, W. et al. 1989*a*). However, washing (by immersion) of sea buckthorn fruits in chlorinated water containing a surface active agent prior to pressing juice caused a decrease in the soluble solids from 10.2 ± 0.349 °Brix to 8.40 ± 0.245 °Brix. (Beveridge 2002) and these two averages are significantly different (*P*< 0.05). This effect is almost certainly due to the tearing of the skin described earlier and the exchange of soluble solids and water that is promoted by this tearing. This offers potential difficulties, since unless the microbial loads in the wash water are carefully controlled, the possibility of spoilage organisms infecting the washed fruits and being carried over to juice products made from these washed fruits is quite high.

The possible contamination of the juice with pathogens present in the wash water must also be considered. Probably inspection of the fruit to remove obvious diseased fruits and leafy debris follow by immediate juice extraction and pasteurization of the juice is required to assure a product safe for consumption. Considering that the skin damage seems to be primarily at the point where the pedicel emerges from the fruit, it might be advantageous to wash with spray water instead of immersion, since this would minimize contact between that part of the fruit that is compromised and the water. Also, some of this problem might be alleviated if an abscission layer could be induced in the fruit to seal the fruit where the stem is detached. Neither of these latter ideas have been tested experimentally at this time.

Considering these difficulties inherent in sea buckthorn harvesting it is clear that the fruits must be transported as quickly as possible to the processing plant. There they must be cooled immediately to temperatures around 4–6°C or less to retard growth of microorganisms. This cooling is best done in circulated cool air storage since the fruits will probably pick up significant water if hydro-cooled. If cooled air is used as the heat transfer medium, then perforation of the baskets holding the fruit to allow maximum air circulation around the fruit should be considered. If the fruits are to be stored more than a day or two they should be stored frozen, preferably frozen by individual quick freezing techniques as this will allow long-term storage, and the fruits may be thawed and processed to products as required on demand. Juice extracted by pressing or centrifugal techniques must be stored refrigerated and should be pasteurized and frozen for long-term storage. Alternatively they may be processed into pasteurized or sterilized finished products and stored in that form at room temperature. It should be remembered that even in this finished state, the shelf life is limited, depending on the product, and the warehousing period should be as short as is practicable to assure delivery of high quality products to the market. Shelf life is improved by cool or refrigerated storage.

Chapter 11. Chemical Composition and Some Physical Properties

Thomas H.J. Beveridge

Agriculture and Agri-Food Canada, Pacific Agri-Food Research Centre
Summerland, British Columbia, Canada V0H 1Z0

The fruit of the sea buckthorn plant weighs between 270 and 480 mg and averages 350 mg depending upon variety and maturity. Pressing these fruits yields 60–85% juice, a yield of 67% has been reported derived from centrifugal methods (Heilscher and Lorber 1996*a*). The composition of fruits or juice is detailed in Table 11.1 (Beveridge et al. 1999). The fruit is notable in that it contains considerable oil as an inherent part of the fruit and this has a marked effect on the processing of the fruits as will be noted in the Chapter 12. There are two sources of oil in sea buckthorn fruit, the seed which contains 6.47–20.2% (w/w) oil (Table 11.3), but usually between 10 and 15%, and the oil held in the pulpy fruit parts surrounding the seed which is termed pulp oil. If a single fruit is considered to weigh 350 mg and a seed 16 mg and if the juice yield is 73%, then it is a straight forward calculation to suggest that there is 2.44–4.88 mg pulp oil in the extracted juice if the oil content is 1–2% (Table 11.1). On the other hand the seed weighs about 16 mg and contains 1.6–2.4 mg oil. Clearly, on a quantitative basis, there is much more pulp oil in a fruit than seed oil.

Physical properties including specific gravity, conductivity, surface tension, and refractive index are provided. The refractive index values given represent sugar contents of 10.8–15.6% sucrose (average 13.5%), approximating 10.8–15.6°Brix (Triebold and Aurand 1963), in line with the °Brix listed later in Table 11.1. The content of soluble sugars determined refractometrically as °Brix ranges from 9.3 to 22.74 °Brix for sea buckthorn juice, representing a wide range of values. This may be due to the fact that the time of harvesting the fruits, in the fall when it matures or in the winter after it has frozen, provides for a harvesting component to the normal seasonal variation of composition (Agafonova and Borodachev 1986). Measurements on fresh ripe fruits grown in Saskatchewan gave levels ranging from 9.7–10.8 °Brix for the variety Indian-Summer (Beveridge et al. 2002). The sugar content is mostly glucose and fructose with minor amounts of xylose and the sugar alcohols; mannitol, sorbitol, and xylitol. This will have only a slight effect on the °Brix values reported, since the glucose, fructose, and sucrose have very similar refractive indices at equal concentrations (Triebold and Aurand 1963).

The value of sea buckthorn is often based on the nutritional value of its fruit. Vitamin C (ascorbic acid) represents a nutrient of major importance in sea buckthorn juice because

Table 11.1. Composition of sea buckthorn fruits/juice.

Attribute (Units)	Range	Average	Identification/Var.	Reference
Fruit Weight (mg)	270–480	350	cv. Indian-Summer	Li et al. (2002)
Fruit Moisture Content (%)	73.6–85.3[a]	82.3	cv. Indian-Summer	Li et al. (2002)
	72.2–75.5[b]	74.2	subsp. *sinensis*	Ma et al. (1989)
	61.5–79.4[b]	70.5	subsp. *sinensis*	Zhang, W. et al. (1989*b*)
Juice Oil Content (%)	0.26–1.43	0.903	subsp. *sinensis*.	Zhang, W. et al. (1989*b*)
	1.8–2.9 (pulp)	2	subsp. *sinensis*	Ma et al. (1989)
Specific Gravity	1.0275–1.0454	1.03797	subsp. *sinensis*.	Tong et al. (1989)
Conductivity (μΩ/cm)	0.297–0.539	0.359	subsp. *sinensis*.	Tong et al. (1989)
Surface Tension (N/m)	46.23–55.14	50.74	subsp. *sinensis*	Tong et al. (1989)
Refractive Index	1.3491–1.3566	1.3533	subsp. *sinensis*.	Tong et al. (1989)
Total Carotenoid (mg/100 g)	9.4–34.5	16.9	cv. Indian-Summer	Li et al. (2002)
	4.6–12.0	7.6	subsp. *sinensis*.	Ma et al. (1989)
	2.0–16.1	6.33	subsp. *sinensis*.	Zhang et al. (1989*b*)
	16–28	–	–	Li and Schroeder (1996)
Vitamin C (mg/100 g)	502–1061	709	subsp. *sinensis*.	Ma et al. (1989)
	360–2500	–	–	Li and Schroeder (1996)
	1348 (single value)[d]	–	subsp. sinensis.	Liu and Liu (1989)
	513–1676	1038	subsp. *sinensis*.	Zhang, W. et al. (1989*b*)
Soluble Sugars (EBrix)	9.3–17.3	11.4	cv. Indian-Summer	Li et al. (2002)
	10.83–15.55	13.51	subsp. *sinensis*.	Tong et al. (1989)
	10.19–22.74	15.98	subsp. *sinensis*.	Zhang, W. et al. (1989*b*)
	6.4–12.7 (reducing sugar)	9	subsp. *sinensis*	Ma et al. (1989)
Glucose (% of total)	49.5–62.1	54.2	subsp. *sinensis*	Ma et al. (1989)
Fructose (% of total)	37.3–50.4	45.4	subsp. *sinensis*	Ma et al. (1989)

Table 11.1. Composition of sea buckthorn fruits/juice. *(Concluded)*

Attribute (Units)	Range	Average	Identification/Var.	Reference
Mannitol (µg/g)	174	na	Finnish Sea buck.	Makinen and Soderling (1980)
Sorbitol (µg/g)	36164	314	Finnish Sea buck.	Makinen and Soderling (1980)
Xylose (% of total)	0.1–0.7	0.42	subsp. *sinensis*	Ma et al. (1989)
Xylitol (µg/g)	15–91[e]	39.2	Finnish Sea buck.	Makinen and Soderling (1980)
Xylose (µg/g)	13–100[e]	45.5	Finnish Sea buck.	Makinen and Soderling (1980)
Organic Acid (% malic)	3.5–4.4[c]	4	subsp. *sinensis*	Ma et al. (1989)
	4.61–7.35[c]	6.05	subsp. *sinensis*	Zhang, W. et al. (1989b)
Malic Acid (%)	1.11–2.34 (L-malic)	1.85	subsp. *sinensis*	Ma et al. (1989)
	2.82–6.08	4.57	subsp. *sinensis*	Zhang, W. et al. (1989b)
Citric Acid (%)	0.042–0.234	0.111	subsp. *sinensis*	Ma et al. (1989)
Tartaric Acid (%)	0.013–0.014	0.0135	subsp. *sinensis*	Ma et al. (1989)
Succinic Acid (%)	0.236–0.643	0.474	subsp. *sinensis*	Ma et al. (1989)
d-Malic Acid (%)	0.015–0.054	0.033	subsp. *sinensis*	Ma et al. (1989)

[a]Dried Weight; [b] Press Juice; [c] Determination as malic acid is a reasonable assumption, but is not stated explicitly; [d] Single value, unripe fruits; [e] Varies with maturity.

it is present in large quantities (Tables 11.1, 11.2), ranging from about 105–2500 mg/100 mL. Considering that orange juice provides 35–56 mg/100 mL (Araujo 1977), the value of sea buckthorn as a new and important source of vitamin C is apparent. Carotenoid pigments are primarily associated with the suspended solids and pulp oil retained by an ultra filtration membrane (Table 11.2) and are reported to range from 2 to 34.5 mg/100 mL juice (Table 11.1, 11.2). The variation associated with this wide range of values is probably dependant on the method of juice production, because it would be expected that the levels of suspended solids would also vary with production technique. Nevertheless the high levels of vitamin C and carotenoid confer the unique nutritional benefit provided by sea buckthorn juice. If pulp oil can be incorporated into the juice or drinks prepared from juice, carotenoid levels can be enhanced and vitamin E (tocopherols) can be added to the unique composition of the juice (Table 11.3).

The juice is very high in organic acids as reflected in the high levels of titratable acidity, and in the low pH of the juice near 2.7 (Table11.1, 11.2). Quantitatively the most important organic acid reported is malic acid, as well as several other minor acids. A

Table 11.2. Composition of the retentate and permeate following separation from sea buckthorn fruit raw juice using cellulose acetate membranes (from Bock et al. 1990).

Characteristic	Raw Juice	Retentate	Permeate
Dry substance (%)	6.5	16.8	5.25
pH	2.7	2.7	2.7
Lipid (%)	0.83	7.9	0
Protein (%)	0.8	4.18	0.37
Acid (tartaric acid) (%)	4.22	4.2	4.15
Reductive sugar	0.7	–	0.72
β-carotene (mg %)	2.1	21.62	0.004
Vitamin C (mg %)	105.3	109.2	70

[a]Following inversion as glucose.

Table 11.3. Some characteristics of sea buckthorn seed and oil.

Attribute (units)	Range	Average	Reference
Seed Weight (mg)	11-24	16	Li et al. (2002)
Seed Moisture (%,w/w)	5.43-21.9	11	Li et al. (2002)
Seed Oil (%,w/w)	9.69-20.2	14.2	Li et al. (2002)
	7.4-9.9	8.4	Ma et al. (1989)
	8-12	–	Li and Schroeder (1996)
	6.47-10.5	8.76	Zhang et al. (1989*b*)
Carotenoid Content (Oil).			
(mg/100 g)	314-2139	1167	Zhang et al. (1989*b*)
From seed oil	50-85	–	Mironov et al. (1989)
From pulp oil	330-370	–	Mironov et al. (1989)
From seed coat oil	180-220	–	Mironov et al. (1989)
From seed oil	trace	–	Mironov et al. (1989)
From pulp oil	900-1000	–	Mironov et al. (1989)
Vitamin E (mg/100 g)	40.1-103.0	64.4	Ma et al. (1989)
From seed oil	61-113	92.7	Zhang, W. et al. (1989*b*)
From juice oil	162-255	216	Zhang, W. et al. (1989*b*)
From residue	390-540	481	Zhang, W. et al. (1989*b*)

Table 11.4. Elemental composition of sea buckthorn fruits/juice.

Element	Range (µg/mL)	Average	Reference
Potassium	100–806	497	Tong et al. (1989)
	0.147–0.209	0.168	Zhang, W. et al. (1989*b*)
Calcium	64–256	143	Tong et al. (1989)
	93.9–173	113	Zhang, W. et al. (1989*b*)
Phosphorus	82.1–206	131	Zhang, W. et al. (1989*b*)
Magnesium	39.8–103	70.4	Zhang, W. et al. (1989*b*)
	53.3–165	88.9	Tong et al. (1989)
Sodium	17.7–125	76.9	Zhang, W. et al. (1989*b*)
	18.0–89.8	48.5	Tong et al. (1989)
Cobalt	< 0.1	-	Zhang, W. et al. (1989*b*)
	0.01–0.09	0.034	Tong et al. (1989)
Chromium	0.108–0.287	0.178	Zhang, W. et al. (1989*b*)
	0.47–1.00	0.699	Tong et al. (1989)
Copper	0.158–0.653	0.384	Zhang, W. et al. (1989*b*)
	< 10	-	Liu and Liu (1989)
Manganese	1.17–2.60	1.67	Zhang, W. et al. (1989*b*)
	0.81–3.86	1.27	Tong et al. (1989)
Nickel	0.115–0.357	0.237	Zhang, W. et al. (1989*b*)
	0.39–0.09	0.189	Tong et al. (1989)
Strontium	0.19–0.616	0.429	Zhang, W. et al. (1989*b*)
	0.08–0.45	0.195	Tong et al. (1989)
Vanadium	0.002–0.009	0.0069	Zhang, W. et al. (1989*b*)
Iron	4.13–10.9	7.58	Zhang, W. et al. (1989*b*)
	5.93–161	28.2	Tong et al. (1989)
Molybdenum	0.03–0.058	0.044	Zhang, W. et al. (1989*b*)
	1.18	1.18	Tong et al. (1989)
Zinc	0.431–1.25	0.763	Zhang, W. et al. (1989*b*)
	2.09–6.31	3.29	Tong et al. (1989)
Tin	0.045–0.259	0.17	Zhang, W. et al. (1989*b*)
Selenium	7.96–11.3	9.21	Zhang, W. et al. (1989*b*)
	0.94–1.11	1.02	Zhao et al. (1989)
Boron	0.43–1.38	1.06	Zhang, W. et al. (1989*b*)
Barium	0.168–0.362	0.244	Zhang, W. et al. (1989*b*)
Aluminium	2.2–16.7	7.88	Zhang, W. et al. (1989*b*)
Titanium	0.103–0.814	0.407	Zhang, W. et al. (1989*b*)
Lithium	0.132–0.303	0.203	Zhang, W. et al. (1989*b*)
	0.06–0.15	0.09	Tong et al. (1989)
Cadmium	< 0.05	<0.05	Zhang, W. et al. (1989*b*)
	0.002–0.015	0.0048	Tong et al. (1989)
Arsenic	< 0.5	<0.5	Liu and Liu (1989)
Lead	0.431–0.761	0.551	Zhang, W. et al. (1989)
	< 1	<1	Liu and Liu (1989)
	0.06–0.27	0.01	Tong et al. (1989)

Table 11.5. Amino acid content of chinese sea buckthorn according to Zhang et al. (1989*b*).

Amino acid	Level (mg/100 g)
Aspartic acid	426.6
Proline	45.2
Ammonia	41.8
Threonine	36.8
Serine	28.1
Lysine	27.2
Valine	21.8
Alanine	21.2
Phenylalanine	20
Glutamine	19.4
Isoleucine	17.4
Glycine	16.7
Histidine	13.7
Tyrosine	13.4
Arginine	11.3
Cysteine	3.3
Methionine	2.3

recent report (Beveridge et al. 2002) documented the importance of quinic acid in the juice derived from the cv. Indian-Summer, ranging from 22.2 ± 1.16 to 26.5 ± 0.09 mg/mL, about twice the malic acid levels.

The protein levels (Table 11.2) are fairly high for a fruit juice and this probably reflects the fact that sea buckthorn juice is a cloudy or opalescent product (Beveridge 2001). The source of the opalescence in most juices is due to the presence of some cell debris, but mostly due to the presence of cellular membranes containing considerable protein and conferring a stable turbidity to the juice (Beveridge 2000). There are many elements and trace elements in sea buckthorn as listed in Table 11.4. Bounous and Zanni (1988) found that fruit maturity affects N, Ca, K, Na, Mg, Cu, Fe, Zn, and Mn contents. Harju and Ronkainen (1984) reported that the elements and trace elements found in liqueurs prepared from sea buckthorn included Al, As, Ca, Cd, Cr, Cu, Fe, K, Mg, Mn, Na, Ni, Pb, Rb, and Zn. Sea buckthorn also has a large number of free amino acids (Table 11.5) of which aspartic acid is, by far, quantitatively, the most important. Each fruit contains a seed weighing, on average, 16 mg, containing 11.0% (w/w) moisture and 14.2% (w/w) oil (Table 11.3). It was reported that fruit maturation affects compositions. Ascorbic acid decreased from 1.48 to 1.1 g/kg between sampling 19 days apart. Quercetin decreased whereas caempferol concentration increased during fruit maturation (Jeppsson et al. 2000).

Table 11.6. Fatty acid composition of oil triacylglycerols of sea buckthorn (*Hippophae rhamnoides*).

| Fatty acid [b] | (mol %) | Seed oil | | Pulp oil | |
| | | Johansson et al. (1987a) | Mironov et al. (1989)[a] | Mironov et al. (1989) | |
		Caucasus (%)	Pamirs (%)	Caucasus (%)	Pamirs (%)
14:0	0.1				
15:0	0.2				
16:0	6.8	20	17	38	38
16:1 (*n*-7)	0.4	5	20	14	50
18:0	1.8	3	1		
18:1 (*n*-9)	12.7	13	19	33	12
18:1 (*n*-7)	2.7				
18:2 (*n*-6)	34.7	40	29	15	1
18:3 (*n*-3)	38.5				
20:0	0.5				
20:1 (*n*-9)	0.2				
20:2					
20:3					
22:0	0.1				
Others	1.4				

[a] Seed lipids contain about 8% of 7-oxononanic acid (see Degering 1963 for nomenclature) and lipids claimed to contain vitamins A, E, and P.
[b] Nomenclature (e.g., *n*-9) indicates the position of the first double bond from the methyl carbon (methyl carbon = 1).

Sea buckthorn oils vary in their vitamin E content, depending on whether derived from seed oil (64.4–92.7 mg/100 g seed), juice oil (216 mg/100 g fruit) or from the pulp after juice and seed removal (481 mg/100 g fruit) (Table 11.3). Carotenoids (Table 11.3) also vary, depending upon the source of the oil. Mironov et al. (1989) indicated the carotenoids to consist of ~20% β-carotene, ~30% γ-carotene, ~30% lycopene as well as a further 15 oxygen containing carotenoids. The tocopherol (vitamin E) fraction was made up of ~50% α-tocopherol, ~40% β-tocopherol and ~10% γ-tocopherol (Mironov et al. 1989). The fatty acid composition of the seed and pulp oils is shown in Table 11.6 in either mol% (Johansson et al. 1997a) or % (Mironov et al. 1989) terms. The seed oils are highly unsaturated, with up to 73% or more of the fatty acids making up the oil being either linoleic or linolenic acids (Lu 1992). Pulp oil is more saturated, with about 38% of the fatty acids making up this oil being palmitic acid, and 14–50% of the fatty acids being palmitoleic acid. The difference between seed and pulp oil seems to lie in the relatively high content of C_{16} fatty acids in the pulp oil and the relatively high proportion of C_{18}

Table 11.7. Phytosterol in Chinese sea buckthorn fruit and seed oil of *H. rhamoides*, spp. *sinensis* (from Yaonian et al. 1995).

Compound	Fruit oil	Seed oil			
		Pressed	Hexane	CO^2	FCHC
Total Sterol					
(mg/100g)	770.6	193.6	1298.4	1217.1	1022.1
Cholesterol (%)	0.3	0.2	0.2	0.1	0.4
Campersterol (%)	3.2	2.16	2	1.9	2.2
7.25 Stigmasterol (%)	1.3	0.3	0.3	–	–
Sitosterol (%)	84.9	72.6	72.1	74.8	73.9
7-Sitosterol (%)	2.7	0.7	0.7	0.7	0.6
5-Avenasterol (%)	5.1	22.5	23.2	19.7	19.3
7-Avenasterol (%)	2.5	–	1.5	1.1	1.6

CO_2; Supercritical carbon dioxide extracted; FCHC; Fluorinated and chlorinated hydrocarbon extracted; – not reported.

Fig. 11.1. Differential Scanning Calorimetry (DSC) of sea buckthorn oil. Sea buckthorn seed oil represents a stored sample and the oxidized sample has been purposely oxidized by bubbled air.

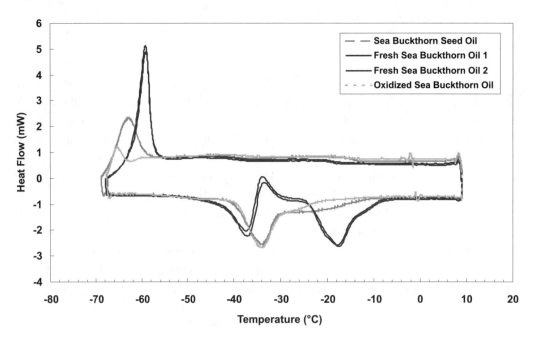

fatty acids in the seed oil. Franke and Müller (1983a) analysed the fat of pulp and seeds and found 47 and 21% saturated fatty acids and 53 and 39% unsaturated fatty acids, respectively. These numbers agree with the data in Table 11.6 in general and emphasize the differences between pulp and seed oils in terms of unsaturation, but also emphasizes the unsaturated nature of the oils. The next most important fatty acid is oleic at 13 –19% (seed oil), and 12–33% (pulp oil), underlining the highly unsaturated nature of the oils in both the pulp and seed (Table 11.6) (Mironov et al. 1989). The seed oil composition described by Johansson et al. (1997a,b) was obtained by extraction with chloroform:methanol (2:1,v:v) mixture, and cleaned up with florisil extracted with hexane:diethylether (4:1,v:v), and measured as the methyl esters gas chromatographically. Negative ion chemical ionization mass spectrometry confirmed the dominance of the C_{18} fatty acids in the fatty acid profile of sea buckthorn seed oil (Johansson 1997a,b). They also reported a small amount of the unusual oleic acid isomer having the double bond 7 carbons from the methyl carbon, while Mironov et al. (1989) reported 8% 7-oxononanic acid. The presence of these unusual acids in sea buckthorn oil await confirmation by future analyses. An unsaponifiable fraction was obtained (1–2%) for seed oil and 0.3% for pulp oil (Miranov 1989).

Phytosterols are plant sterols with structures related to cholesterol, which are capable of lowering plasma cholesterol on consumption by humans. Elevated blood cholesterol is one of the well established risk factors for coronary heart disease and lowering this indicator can presumably impact heart disease incidence (Anonymous 2000; Thurnham 1999). Phytosterols are the major constituents of the unsaponifiable fraction of sea buckthorn oils. As can be seen from Table 11.7, the major phytosterol in seabuckthorn oil is sitosterol (β-sitosterol), with 5-avenasterol second in quantative importance. Other phytosterols are present in relatively minor quantities. The total quantities of phytosterol is quite high in sea buckthorn and may exceed soybean oil by 4–20 times. Clearly, as a source of dietary sterols, sea buckthorn is worthy of further consideration.

Some general properties of sea buckthorn seed oil are given in Table 11.8. The seed oil is yellow in color and absorbs very strongly in the UV-B range, which makes it useful as a natural sun screen. The high conjugated dienes, ρ-anisidine, and peroxide values suggest low stability to oxidation consistent with the high level of unsaturation displayed by the fatty acid composition. The saponification value is comparable to that of more common vegetable oils. The distribution of tocopherols does not agree particularly well with the determinations of Mironov et al. (1989), except that α-tocopherol is about 50% of the total. The high vitamin E equivalency provides protection against oxidation and makes the oil a useful natural antioxidant.

Sea buckthorn seed oil has unique melting and crystallizing characteristics (Fig. 11.1). Fresh sea buckthorn seed oil crystallizes at −59°C, a temperature much lower than typical vegetable oils, such as canola (−43°C) and sunflower (−45°C) oils. Two polymorphic forms, high melting and low melting, with endotherms at −17.5 and −37.5°C, respectively, characterize the melting behavior. Storage and oxidation result in deterioration of

Table 11.8. Physicochemical characteristics of sea buckthorn seed oil, cv. Indian-Summer.

Characteristic	Mean Value	Std. Deviation
Absorptivity (L/g*cm)		
232 nm	2.89	0.03
270 nm	0.64	0.02
303 nm	0.41	0.02
410 nm	0.06	0.02
Diene value	3.16	0.01
Triene value	0.07	0.002
ρ-Anisidine value	34.19	0.06
Peroxide value (meq/kg)	20.68	0.06
Saponification number	190	1.63
Viscosity (mPas.s)	44	0.5
Carotenoid content (mg/100 g)	41.1	13.4
Tocopherol content (mg/100 g)		
alpha	155	7
beta	16.4	1.7
gamma	134.9	2.8
delta	11.3	1.4
Vitamin E equivalent (mg/100 g)	175	8

the oil characterized by lowering of the crystallization temperatures towards the −62 to −65°C range (Fig. 11.1). A single melting point at about −34.2°C was observed for oxidized sea buckthorn seed oil.

Chapter 12. Processing and Products

Thomas H.J. Beveridge

Agriculture and Agri-Food Canada, Pacific Agri-Food Research Centre
Summerland, British Columbia, Canada V0H 1Z0

There is a wide array of products possible from sea buckthorn fruit, as wide as the lines of products generally available from any other fruit. Many claims are made as to the efficacy of these products for use as food, fresh fruit, health foods or nutraceuticals, pet foods, cosmetics, and skin preparations for improving the health and appearance of the skin. As indicated in Chapter 10 on harvesting and storage, the skin of the fruit tends to be damaged at the pedicel end, causing weeping following harvest. This makes production of fruit for the fresh market difficult. Nevertheless, some use of fresh product has been proposed as additions to or garnishes supporting the appearance and flavor of main courses. The sea buckthorn fruit consists of a relatively tough skin surrounding a pulpy liquid which in turn surrounds a single-sheathed seed. A typical diagram describing sea buckthorn processing is given in Fig.12.1. The most obvious procedure to begin the separation process is to apply a press to extract juice from the fruit.

Juice Extraction

Washing of fruits prior to juice extraction runs the risk of soluble solids dilution by uptake of the wash water as discussed in Chapter 10, so washing by immersion of the fruits in the washing medium is probably not advisable. Prior to further processing the fruits should be carefully inspected and diseased fruit, debris resulting from mechanical picking, such as leaves and stems, and dirt or orchard debris removed before juice extraction. Further, it is important that the fruits be either newly picked or refrigerated immediately after picking. Presses for fruit juice extraction are supplied in various forms, such as serpentine belts, rack and cloth, or screw presses as described extensively by Bump (1989). One would anticipate that a serpentine belt press or a rack and cloth press would suffice, since trials with rack and cloth presses indicate that these presses work reasonably well (Fig. 12.2). One of the problems associated with pressing sea buckthorn fruits is that all the fruits do not burst and release juice during pressing. In this respect, a serpentine belt press in which the fruits are placed between two belts in a thin layer and then pressed between rollers may work better than a rack and cloth type of press. Otherwise, one would be well advised to provide means at the front end of the process for bursting the fruits prior to pressing. A pair of counter rotating drums close set below the diameter of the fruit will suffice for this purpose (Fig. 12.3).

Fig. 12.1. Diagrammatic outline of the procesing of sea buckthorn fruits.

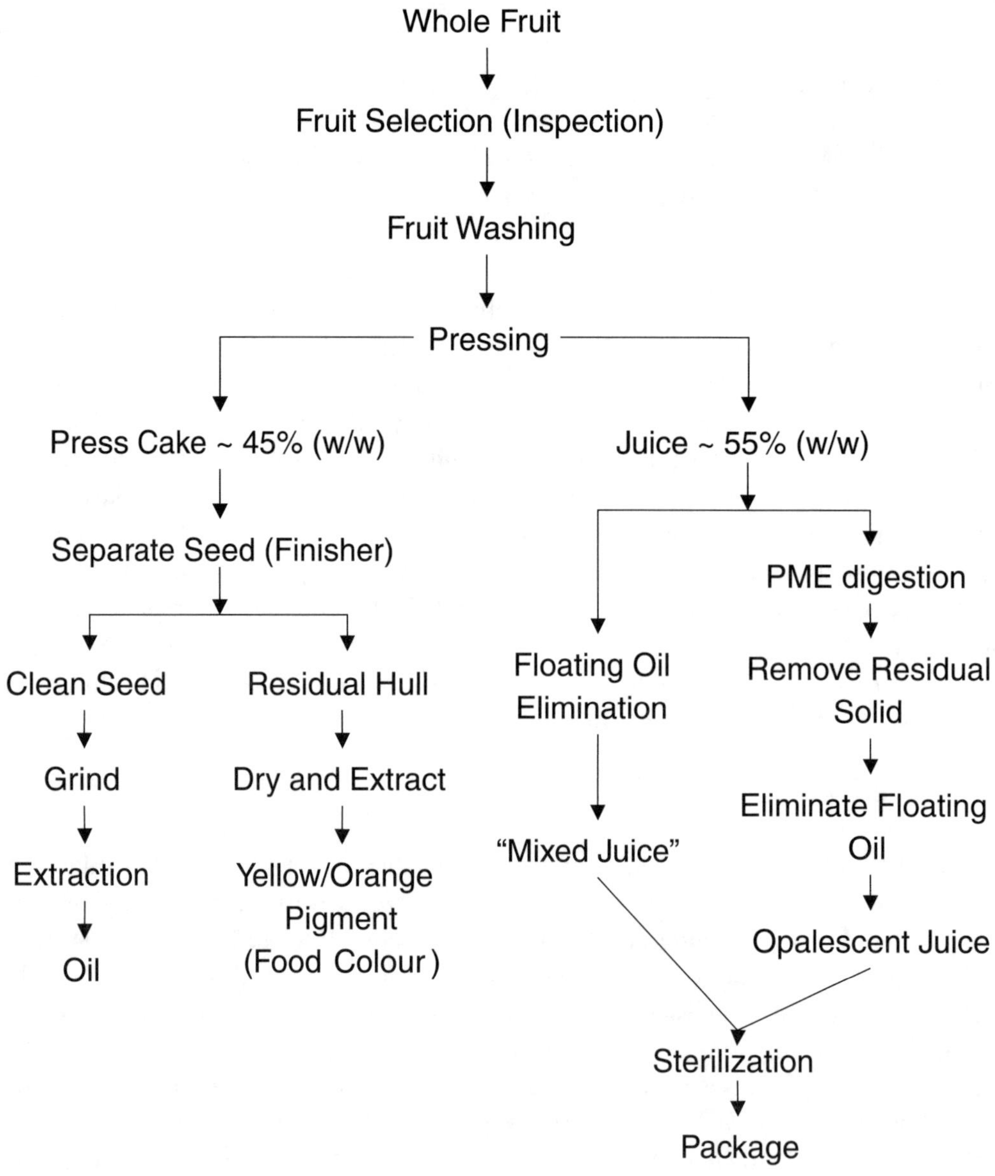

```
                            Whole Fruit
                                 │
                                 ▼
                    Fruit Selection (Inspection)
                                 │
                                 ▼
                          Fruit Washing
                                 │
                                 ▼
        ┌──────────────────── Pressing ────────────────────┐
        │                                                  │
        ▼                                                  ▼
Press Cake ~ 45% (w/w)                           Juice ~ 55% (w/w)
        │                                       ┌──────────┴──────────┐
        ▼                                       │                     ▼
Separate Seed (Finisher)                        │              PME digestion
   ┌────────┴────────┐                          ▼                     ▼
   ▼                 ▼                  Floating Oil          Remove Residual
Clean Seed      Residual Hull            Elimination               Solid
   │                 │                          │                     ▼
   ▼                 ▼                          ▼              Eliminate Floating
 Grind         Dry and Extract           "Mixed Juice"               Oil
   │                 │                                                ▼
   ▼                 ▼                                         Opalescent Juice
Extraction     Yellow/Orange
   │              Pigment
   ▼           (Food Colour)
  Oil
                                              Sterilization
                                                    │
                                                    ▼
                                                 Package
```

Sea buckthorn juice obtained by pressing is a complex mixture of large and small suspended solid particles and droplets. creating a three-phase system containing up to 2.9% oil (Beveridge et al. 1999). The juice is very turbid (Fig. 12.4). Microscopically, the oil droplets appear as round, distinctly orange spheres staining red-orange with Oil O Red (white arrows, Fig. 12.5; Oomah et al. 1999) and the particulate matter is translucent or granular in appearance (black arrows, Fig. 12.5). The orange color of the oil droplets is probably due to their high carotenoid content. Many large aggregates of cellular material

Fig. 12.2. Press cloth contents after pressing for 5 minutes at 3000 psi showing the proportion of unbroken fruits. One can also notice the leaves and debris present in these hand picked fruits.

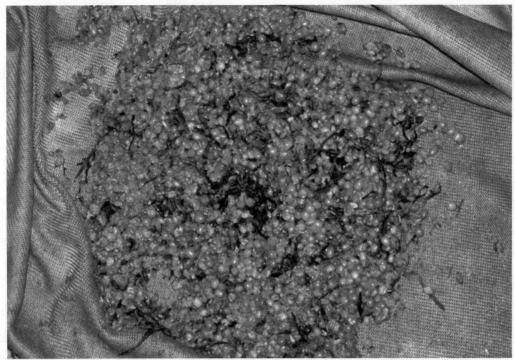

are yellow to dark orange in the original juice and are clearly visible in Fig. 12.5 (black arrows). Heating to 90°C for 3 minutes. causes an obvious increase in the juice consistency observed as reduced flow rates from containers of heated juice. These changes are reflected in the microscopic morphology of the juice particulate (Fig. 12.5). Some oil drops increase considerably in size as expected because coalescence of oil droplets is enhanced by increased temperature. Large droplets grow at the expense of smaller droplets, resulting in a lower total interfacial surface energy in the system and a wider range of droplet sizes. The granular particulate matter changes through either aggregation of the condensed particles or disruption of the compact structures to porous more open structures. The yellow-orange pigment noted in Fig. 12.5 as localized primarily in the granular particulate has been dispersed throughout the juice in Fig. 12.5, supporting a contention that the particulate has been dispersed or solubilized by the heating procedure, at least to some extent.

If freshly pressed juice is allowed to stand for 1 or 2 days, it will separate into three phases: a floating, oil-containing phase in the upper third of the tube (Fig. 12.4), a liquid, opalescent portion in the center of the tube, and a sinking particulate sediment. This separation is claimed to be undesirable from a consumer's point of view (Kleinschmidt et al. 1996). The separation behavior is similar to raw milk and can be handled in the same way. Vertical axis, high speed centrifuge of the disk stack type (Fig. 12.6; Beveridge 2000), which operate similarly to a cream separator, will allow rapid separation of the fat rich and

Fig. 12.3. Prototype roller machine for crusing sea buckthorn fruits. The feed hopper and protective cage have been removed for clarity. Dual drives for differential roller speeds are not required.

aqueous phases (Zhang, W. et al. 1989*a*). If pulp oil is left in the juice, it will result in the formation of an oil layer on the juice surface, creating an oil ring that remains on the package surface after the juice is removed. This oil ring remaining on the package is unsightly and undesirable, and centrifugally reducing the juice oil content below about 0.1% will eliminate the floating oil problem. At the same time as the oil is removed by the disk stack centrifuge, the coarse sediment will be sedimented to the bottom of the bowl and can be removed automatically by the desluging mechanisms present in the centrifuge (Fig. 12.6). This centrifugal process results in the production of an oil rich cream phase, which can act as the raw material for extraction of pulp oil, an oil free, opalescent or cloudy aqueous juice, which can be packed and consumed as such after appropriate pasteurization and packaging, and a waste solid sediment, which might provide a source of pigment or of sea buckthorn solids if properly encapsulated for nutraceutical use or for use as a dietary supplement.

Considering centrifugal technologies further, the use of decanter (horizontal axis) centrifuges (Fig. 12.7) would be a further possibility. These centrifuges have been shown

Fig. 12.4. Sea buckthorn juice obtained by pressing followed by gravity separation overnight. The juices have saparated into three phases, a cream phase above the arrow head, and a small portion of sedimented solids visable as a button on the bottom of tubes 1–4. Sedimented solids were not clearly visible in tubes 5–6, because of the turbidity of the juice phase. The juice of tubes 2–4 were obtained from commercial pectin enzyme treated fruit mashes. this treatment results in a more opalescent, less turbid, juice product.

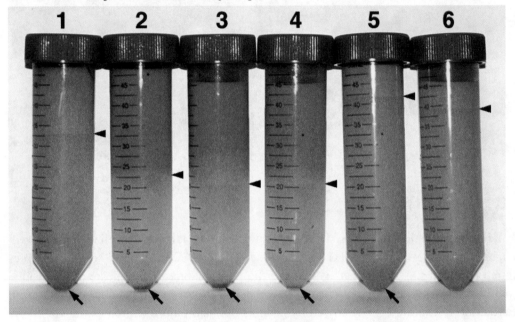

to be useful to extract juice from a wide variety of fruit and vegetable feed stocks (Beveridge 1997), and are applicable to sea buckthorn (Zhang, W. et al. 1989). The use of these machines in place of a press is particularly interesting, since the high shear forces generated in the machine associated with accelerating the mash to bowl speed might make it unnecessary to burst the fruits prior to pumping to the machine. In addition, there are versions of the decanter centrifuge, which are capable of removing suspended solids while simultaneously removing a floating oil layer from the aqueous phase (Beveridge 2000). If this should prove feasible, then it would be possible to produce sea buckthorn juice, cream and sediment directly from fruits with a single machine.

Alternatively, the crushed fruits or extracted juice may be treated with a preparation containing pectin methylesterase (PME) (Lui and Lui 1989), or perhaps treated with one of the many commercially available hydrolytic enzyme preparations. These preparations contain a mixture of carbohydrate hydrolysing activities, including cellulases, pectin methyl esterases and pectinases designed to break down the carbohydrate of the cell wall and middle lamella which cements cells together. Hydrolysis of the middle lamella improves juice release from the tissue by reducing juice viscosity and liquefying part of the cellular material. The character of the suspended solids is also changed, allowing development of a stable cloud in some cases and easy removal of suspended solids in other cases for improved clarification of juices (Beveridge et al. 1999).

Fig. 12.5. Photomicrographs of sea buckthorn juice in transmitted light: (a) fresh, unheated juice; (b) heated juice, 90°C, 3 min. White and black arrows indicate oil droplets and condensed granular material, respectively.

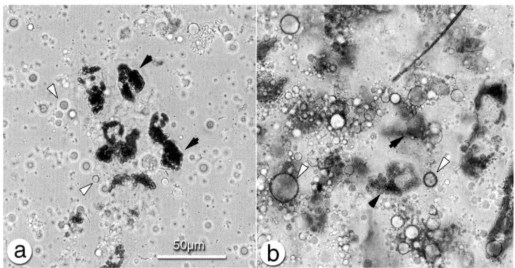

For preservation purposes, it is necessary that the juice be sterilized/pasteurized. High-temperature-short-time (HTST) processes of 80–90°C for several seconds are preferred (Liu and Lui 1989). This is because the juice is somewhat delicate and will give loss of flavor and development of off-flavor if heated beyond the conditions indicated. Furthermore, vitamin C is destroyed by heating, so maximum retention of the vitamin is promoted by HTST conditions.

The juice turns brown after about 6 months at 15–20°C, and this browning is reduced under non-oxidative conditions. Reducing storage temperatures to 4°C prolongs storage life (Zhou and Chen 1989) and enzymes and sunlight are important sources of browning initiation. It is not clear if the browning is due to residual polyphenol oxidase or if the color develops because of nonenzymatic (Maillard or vitamin C) browning, but packaging should limit or eliminate contact of the juice with oxygen and low temperature storage should be considered (Saint-Crieq de Gauljac et al. 1999). Chemical preservation using potassium sorbate (0.45–0.5 g/L) has been reported (Lange et al. 1991).

Normally, sea buckthorn juice is opalescent to very turbid, depending upon how much suspended solid can be left remaining after centrifugation. However, ultra filtration may be used to remove all particulate and produce a clear juice (Bock et al. 1990; Heilscher and Lorber 1996). The ultra filtration membrane can have a molecular weight cut-off of 100 000 or more and the process produces an oil-free permeate and a oil-rich retentate, which can be utilized for production of pulp oil rich in vitamin E, and a solid material rich in carotenoids, which may be used as an isolation source for the pigment or as a dietary supplement.

Fig. 12.6. Disk Stack centrifuge for high speed separation of aqueous and cream layers of sea buckthorn press juice (from Westfalia, with permission).

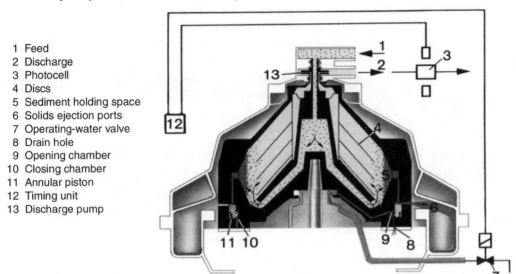

1 Feed
2 Discharge
3 Photocell
4 Discs
5 Sediment holding space
6 Solids ejection ports
7 Operating-water valve
8 Drain hole
9 Opening chamber
10 Closing chamber
11 Annular piston
12 Timing unit
13 Discharge pump

Oil Extraction

Sea buckthorn offers two possibilities for oil extraction. Pulp oil exists in the juice pulp and is isolated as a cream layer by centrifugal technology. This technology is apparently practiced industrially in Germany (Heilscher and Bat 1990). Also, the seeds of sea buckthorn contain an average of 14% (w/w) oil (Beveridge et al. 1999) and this oil is available for extraction. The usual methods for manufacturing oil commercially require counter-current (usually) extraction of the oil bearing material with an organic solvent, commonly hexane (Weiss 1963, 1970). Most organic solvents leave behind a residue of solvent and sometimes this residue may be somewhat toxic. Hexane has the advantage of low toxicity and easy removal from the extracted oil with distillation procedures, and is widely used in commercial extraction of oilseed crops, such as soybean and canola. Hexane is also volatile and flammable, creating potential for hazardous conditions. Increasingly, consumers are demanding fewer residues in their foods and with oils this condition can be met by using newer extraction techniques, such as supercritical fluid extraction (SCE) expecially carbon dioxide under high pressure. Sea buckthorn oil may be amenable to the application of these techniques since it is a specialty oil used in medicinal cosmetics and as a nutraceutical supplement (Beveridge et al. 1999). To provide an economical sea buckthorn oil, the oils, especially the pulp oil, may be extracted into another triglyceride or vegetable oil, such as soybean or olive oil, to form a blended oil. This "sea buckthorn" oil is one form of the "oil-of-commerce" commonly sold from eastern European sources. If solvent residues are of concern, then the source of this extracting oil is of vital importance.

Fig. 12.7. Diagrammatic of a decanter centrifuge showing essential parts. This machine is designed for separation of a liquid and solid phase, but designs exist which allow skimming of a floating oil phase in addition to separation of aqueous and solid phases (from Westfalia, with permission).

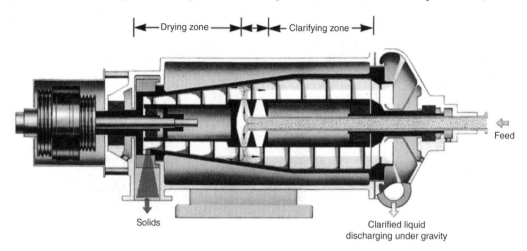

Increasingly, supercritical carbon dioxide extracted oils are becoming available (Heilscher and Bat 1990; Kallio et al. 1995).

A significant advantage of supercritical carbon dioxide extraction over organic solvent extraction is that the solvent can be quickly, easily and completely removed from the accompanying solute, or oil, in the present case. The ideal extractive fluid should be completely inert, nonflammable, inexpensive, have a moderate critical pressure to minimize compression costs, a low to moderate critical temperature to handle thermolabile compounds, a low boiling point for easy removal after extraction (Meireles and Nikolov 1994), and a low viscosity for easy penetration into the material to be extracted. In food and pharmaceutical applications, carbon dioxide meets most of these criteria and is widely used for that reason. The critical point for carbon dioxide (CO_2) with respect to supercritical fluids is shown on Fig. 12.8. For CO_2 the critical point is at 31.05°C and 7.38 MPa (Rizvi et al. 1986) and the particular set of conditions of particular interest is refered to as the region of study.

A diagram of a typical superfluid extractor is shown in Fig. 12.9. In operation the equipment consists of an extraction vessel into which the material to be extracted is placed. This material is commonly finely ground, divided or flaked so as to present maximum surface to the solvent. The equipment consists of a pump to pressurize the CO_2 to the desired pressure after which it is pumped through the filled extractor. After extraction the pressure may be released in stages through several collectors (Fig. 12.9) through a series of valves. Some materials are more soluble in CO_2 at higher pressures, and less soluble at lower pressures. In this way the pressure may be released stepwise allowing isolation of the less soluble components in the series of collectors effecting an approximate separation of components of interest in the extract. The carbon dioxide released from the

Fig. 12.8. Phase diagram for carbon dioxide showing the pressure and temperature conditions for operation as a supercritical fluid (from Temelli et al. 1988).

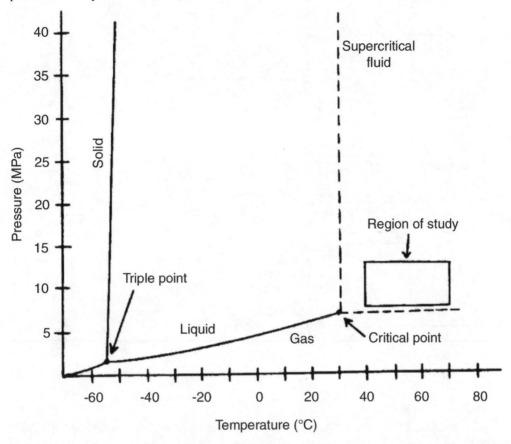

extract is recycled through the pump and reused in a further cycle. Once the material in the extractor becomes exhausted, the extractor vessel is isolated for refilling, and commencement of a new extraction cycle.

Very little data are available on the supercritical extraction of sea buckthorn oils. The report by Stastova et al. (1996) is the only source of direct data for extraction encountered. In referenced work (Stastova et al. 1996), seeds are separated and dried and the remaining pulp containing the pulp oil is air dried. Both materials are ground for extraction with supercritical CO_2. Air dried pulp and seed in mass ratio 45:55 were ground to 0.3 to 0.4 mm, flaked to 0.12–0.4 mm and extracted at 20°C and 5.7 MPa. Recovery was 4–8% (wt/wt), depending on the original oil content of the feed material. In another referenced report, extraction at 35 MPa and 40°C provided oil with yields of 16.5% (wt/wt). In their experimental work, Stastova et al. (1996) indicated that the solubility of sea buckthorn oil in CO_2 at 27 MPa was insensitive to temperature between 25 and 60°C and ranged from about 5.8 to 7.4 mg oil/g CO_2. Pulp oil was about 19% more soluble in supercritical CO_2 than seed oil. Increasing the temperature of extraction from 25 to 60°C increased the rate

Fig. 12.9. Diagrammatic outline of a supercritical fluid extractor showing the essential parts.

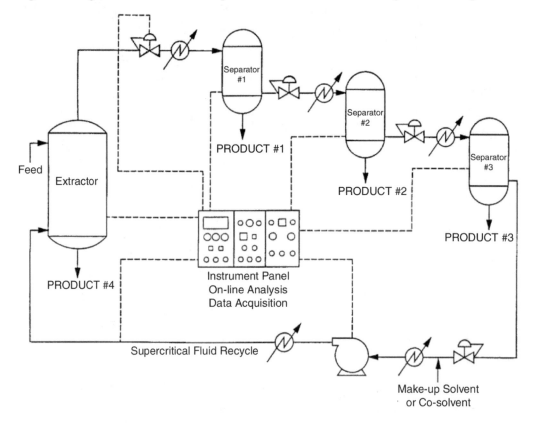

of oil extraction, but not the final amount (Shaftan et al. 1986; Temelli et al. 1988). Increasing pressure from 17.4 to 27 MPa provided similar results. The parameter which most affected the amount of oil extracted was the particle size of the grind, finer grinds provided greater yields. Two periods of extraction were discerned in the extraction time course of sea buckthorn. The first, rapid extraction depends upon the grade of grinding but the second is several orders of magnitude slower than the first and rises slowly with increasing temperature. For the intended purpose of sea buckthorn as a medicinal, cosmetic or nutraceutical ingredient, extraction by supercritical CO_2 is highly desirable and the conditions required for the oil extraction, particularly methods of extracting very finely ground material, is necessary. Also, the pulp oil can be isolated as a cream layer centrifugally and procedures for extraction of this oil enriched, liquid material by supercritical methods would be desirable.

Pigment

A pigment termed "sea buckthorn yellow" can be extracted from sea buckthorn waste material. The waste material could be the press cake remaining after juice extraction or the sediments remaining after centrifugation. In one process the pigment is extracted by low concentrations of alcohol (Chen et al. 1995; Liu and Liu 1989) after adjustment of the suspension to 11–13°Brix by concentration. The extracted pigment is dried by spray drying to yield a yellow powder which is soluble in water, alcohol, acetone, and several petroleum derived solvents. It contains flavones, but also carotene and vitamin E. The pigment absorbs maximally at 213, 315, and 445 nm (Lui et al. 1989) or at 450 nm in aqueous solution (Chen et al.1995). The color is stable to acid, but unstable in base and heating to 80–100°C causes slow degradation. Acute toxicity testing in mice indicated no acute problems associated with the pigment extract. Supercritical CO_2 has also been used to extract a yellow coloring material from sea buckthorn waste. Pressure had the greatest influence on extraction, with yields increasing with extraction pressure. A yield of total carotenoid of 64% was achieved under processing conditions of 60 MPA, 85°C (Messerschmidt et al. 1993).

Teas

Sea buckthorn leaves contain nutrients and bioactive substances. These include flavonoids (Chen et al. 1991), carotenoids, free and esterified sterols, triterpenols, and isoprenols (Goncharova and Glushenkova 1996). Currently in Russia, the leaf is used as a source for extraction of vitamins and flavonoids. From air-dried leaves numerous products can be made, but most important are teas and tea powders. In China, young sea buckthorn leaves were processed by withering, rolling, pan-frying, natural moistening, firing, sorting, and packaging, and produced a promising new product. Sea buckthorn tea has been accepted with enthusiasm by the general public.

Chapter 13. Nutritional and Medicinal Values

Thomas S.C. Li, Thomas H.J. Beveridge, and B. Dave Oomah

Agriculture and Agri-Food Canada, Pacific Agri-Food Research Centre
Summerland, British Columbia, Canada V0H 1Z0

Nutritional values of sea buckthorn fruit are based on its known composition (Beveridge et al. 1999) and the relationship of this composition to human nutritional requirements (Magherini 1986). The fruit, including seeds, contains large amounts of essential oils and vitamin C (Centenaro et al. 1977; Novruzov and Aslanov 1983). The composition of the fruits, juice, and oils are covered in detail in Chapters 9 and 11 and reference to those chapters is encouraged. Generally, sea buckthorn fruits are very high in health promoting compounds (Jeppsson et al. 2000). The vitamin C concentration in fruits varies depending on species, geographical location, and physiological maturity (Bernath and Foldesi 1992; Zhou et al. 1991). Ascorbic acid in sea buckthorn species has been reported to vary from 360 to 2500 mg/100 g (Beveridge et al. 1999) from general literature sources, and varied between 28 and 201 mg/100 g in Finland (Yao et al. 1992; Yao and Tiherstedt 1994), 50–330 mg/100 g in Russia (Rousi and Aulin 1977), and 150–340 mg/100 g in Germany (Jeppsson et al. 2000). It varied between 137 and 193 mg/100 mL for juice prepared from sea buckthorn grown on the Canadian prairies (Beveridge et al. 1999). These values are higher than in virtually any of the commonly consumed fruits such as oranges (50 mg/100 g, Lu 1992), strawberries (64 mg/100 g, Gontea and Barduta 1974), or tomatoes (12 mg/100 g, Lu 1992) and is comparable to kiwi fruit (100–400 mg/g, Lu 1992).

Sea buckthorn is also high in protein, especially globulins and albumins (Solonenko and Shishkina 1983), carotene (Kostyrko 1990), fatty acids (Loskutova et al. 1989), and vitamin E (Bernath and Foldesi 1992). Fatty acid and vitamin E contents are higher than in wheat, safflower, maize, and soybean (Lu 1992). The leaves of sea buckthorn contain many nutrients, such as proteins and bioactive substances (Morar et al. 1990), which are especially useful in animal feed.

Sea buckthorn is regarded as the next generation of new botanical because of its considerable medicinal value. Medicinal uses of sea buckthorn are well documented in Asia and Europe. Historically, the Chinese have used sea buckthorn medicinally for thousands of years. Clinical investigations on medicinal uses were initiated in Russia during the 1950s (Gurevich 1956). In 1977, sea buckthorn was officially listed in the Chinese Pharmacopoeia by the Ministry of Public Health and the reputation of sea buckthorn as a medicinal plant was established. More than 10 different drugs have been developed from sea buckthorn and are available in different forms (e.g., liquids, powders, plasters, films, pastes, pills, liniments, suppositories, aerosols, etc.), they can be used to treat oral, rectal, and vaginal mucositis, cervical erosion, radiation damage, burns, scalds, duodenal ulcers, gastric ulcers, chilblains, and skin-ulcers caused by malnutrition and other skin damage (Abartene and Malakhovskis 1975; Buhatel et al. 1991; Chen 1991; Cheng et al. 1990; Dai et al. 1987; Kukenov et al.1982).

Flavonoids, such as leucocyanidin, catechin, and flavonol (isorhamnetin, quercetin, quassin, and camellin), and a trace of flavanone are abundant in sea buckthorn leaves and fruits (Hakkinen et al. 2000). This and other evidence suggest that sea buckthorn can strengthen the immune system, increase human resistance against disease, and postpone senescence (Xu et al. 2001). It may also decrease peripheral vessel resistance to osmotic transfer and increase blood vessel elasticity. Phenols are effective against oxidation, tumorigenesis, and radiation (Chen 1988). A total flavonoid extract from sea buckthorn has anti-myocardial, anti-hyperlipemia, and anti-fat-liver effects (Chai 1989). It was shown to remit angina and improve cardiac rhythm, to improve functioning of the cardio-vascular system, perhaps through a direct effect on the heart muscle, and to treat coronary heart disease (Zhang 1987). In a randomized, double blind, placebo controlled study, Wang et al. (2001) suggested that total flavonoids improved myocardial contractility and strengthened cardiac pump function in normal human subjects. China now produces acetylsalicylic flavonoid tablets as an prescription drug for heart function improvement. Sea buckthorn extracts may strengthen non-specific immunity functions, as demonstrated by anti-anaphylactic effects and increased phagocytic activity, and the juice is reported to improve the multiplication index of splenic lymphocytes in mice (Zhong 1989). Sea buck-thorn juice contains superoxide dismutase, which provides anti oxidant activity, clearing free radicles from membrane systems (Jin 1989).

Sea buckthorn oil is approved for clinical use in hospitals in Russia, and in China where it was formally listed in the "Pharmacopoeia" in 1977 (Xu 1994). In early days, the most important pharmacological functions of sea buckthorn oil could be summarized as antiinflammatory, antimicrobial, pain relief, and promotion of regeneration of tissues (Li and Wang 1998). The anti-inflammatory action of sea buckthorn oil is conferred by its high content of vitamin E, which can eliminate free radicals and decrease oxidation of unsaturated fat in the fatty membrane (Zhen et al. 1996). It also can be used for skin graft-ing, cosmetology, and treatment of corneal wounds. Russian researchers reported that 5-hydroxytryptamine (hippophan) isolated from sea buckthorn bark inhibited tumor growth (Sokoloff et al. 1961). Others reported that the organic extract of sea buckthorn is safe and has adaptogenic immunomedulating properties and improves the mental performance (Agrawal et al. 2001c).

The therapeutic wound-healing effect of sea buckthorn oil has been known for a decade (Kostrikova 1989) and has also been used successfully in the treatment of 350 patients with melanosis, senile skin wrinkles, and freckles (Zhong 1989). Sea buckthorn extracts could improve the micro-circulation through scalp blood capillary vessels, and reduce hair loss. Sea buckthorn oil, which is safe and stable for a long period of time, is used as a major ingredient for the cosmetic cream, since it possesses therapeutic effects on skin (Singh 2001). Agrawal et al. (2001 *a, b*) reported that sea buckthorn is an effective remedial measure for the management of impaired mental functions, particularly to stop further memory loss among elderly people.

Sea buckthorn oil, fed to guinea pigs, showed increased activity of some erythrocyte membrane enzyme systems and a reduced level of malondialdehyde (an oxidation product) in erythrocyte membranes (Rui et al. 1989). It was also reported that the oil can inhibit tumor development of transplanted tumors and both the oil and juice can kill S180 and P388 cancer cells and inhibit strains of human gastric carcinoma (Zhang 1989). It can inhibit the liver cancers induced by aflatoxin B_1 and inhibit the formation of N-nitroso compounds preventing the induction of cancers. Sea buckthorn oil has been shown both in oral administration and topical application, to reduce inflammation, reduce pain and promote repair of tissue in animal models. Most health related data has been developed by using animal models such as induced gastric ulcerations, acute radiodermatitis in rats and other radiation induced injuries (Zhang 1988). Similar effects were found in limited clinical trials. Sea buckthorn seems to prevent lipid oxidation and these antioxidant activities are thought to contribute to the postponement of senility. There is evidence that sea buckthorn contains compounds that are anti-cancer in nature and these compounds are present in both the oil and juice fractions (Xu et. al. 2001). The medicinal benefits of sea buckthorn oil has propelled research in the efficient extraction of oil and components thereof, their characterization and the understanding of the health-related effects of oil and its associated components.

Fruits of sea buckthorn have been used in Tibetan, Mongolian, and Chinese traditional medicines for the treatment of different diseases for more than 1000 years (Zhou and Jiang 1989). Sea buckthorn juice has a very high antioxidant activity, as high or higher than blueberries (Velioglu et al. 1998). The therapeutic activity of the aqueous portion of sea buckthorn fruits may be due to this antioxidant capacity, which is attributed primarily to phenolic levels in the juice with which the activity correlates (Gao et al. 2000). As well, some of the antioxidant activity would be expected to be contributed by the ascorbic acid known to be present in high amounts (Chapter 11). Lipophilic antioxidant activity in sea buckthorn correlates, or is associated, with carotenoid levels in hexane extractives, although contributions from tocopherols (vitamin E) would be expected. Sea buckthorn oil prevents lipid peroxidation in animals more effectively than vitamin E and the strong antioxidant activity of oil extracted from sea buckthorn fruit flesh protects erythrocyte membranes against peroxidation and cell aging in guinea pigs (Rui et al. 1989). Crude extracts of sea buckthorn fruit exhibit higher, metal ion catalysed, peroxidation inhibition than aqueous phenolic extracts. Hence, the anti-aging activity in the oil is postutated to be a component in the oil other than vitamin E, and carotenoid compounds seem probable candidates. However, fruit (pulp) oil of sea buckthorn contains a large number of bioactive substances (Schapiro 1989) and more than 100 compounds have been identified from the unsaponifiable fraction of fruit oil, some of which may contribute to the biological activity of sea buckthorn oil (Wang et al. 1989). Sea buckthorn oil triglycerides contain high concentrations of palmitoleic acid (>17%) that may have cholesterol and triglyceride lowering as well as stroke-suppressing effects, and the phytosterol, sitosterol, present in large amounts in the unsaponifiable fraction, is known to inhibit platelet aggregation. These effects will be noted in some detail in the following discussion.

Topical application of seed and pulp oils on burned, scalded, wounded, or radioactively damaged skin of both humans and experimental animals has shown healing and anti-inflammatory effects (Vlasov 1970; Mironov et al. 1980 1991; Li and Xu 1993; Zhao 1994). It reduces tissue inflammation, accelerates tissue regeneration and has been used to improve healing in 1st, 2nd, and 3rd degree burns and skin grafts. In 79 cases of cervical erosion, the oil was applied daily for 5 days and relieved chronic cervicitis or cervical erosion in 50–74% of the cases observed (Xu et al. 1989). Also, improved healing in 102 cases of radiation induced dermatitis, mucosites and burns were observed. The antiinflammatory effect of the oil is usually considered to be brought about by the presence of ascorbic acid, tocopherol, plant sterols, and especially carotenoids, although evidence suggests that this effect can be enhanced by the triacylglycerol of the oil (Vereshchagin and Tsydendambaev 1995).

In a study assessing the reparative effect on skin wounds in rabbits, sea buckthorn oil stimulated the regeneration of skin and complete healing was achieved in 13–14 days compared to 19–21 days for the control treated with sunflower oil. Gastric ulcers in rats induced by acetic acid shows healing, estimated as ulcer area reductions to 6.0 ± 0.7 mm^2 for sea buckthorn oil administered by esophagus compared to reduction to 21 ± 1.3 mm^2 for the control sunflower oil (Mironov et al. 1989). Animal studies showed that sea buckthorn oil had anti-inflammatory activity by inhibiting the development of the inflammatory reaction in the hypodermic skin tissue of mice (Lebedeva et al. 1989), and appears to suppress the secretion of gastric juice in the gastrointestinal tract of cats and dogs (Khaidarov et al. 1989). It quickly alleviates and cures esophagal and duodenal inflammations in rats. The anti-ulcer constituent, purported to be β-sitosterol-β-D-glucoside, is present at 10 times higher levels in seed oil than in pulp oil (Jiang et al. 1989). Sea buckthorn oil extends by 30%, the survival period of radiation-treated mice by slowing the increase in polychromatic erythrocytes.

Treatment of operative wounds in ear, nose, and throat with sea buckthorn oil has been documented. Sea buckthorn oil has also been used successfully in the clinical treatment of 56 cases of traumatic perforation of the tympanic membrane (Xu et al. 2001). Sea buckthorn oil eases pain, decreases allergic reactions, and facilitates exfoliation of the post-tonsillectomy nick membrane and has therefore been used in the treatment of the postoperative wounds of tonsillitis (Fayman1991). Sea buckthorn cream made essentially from sea buckthorn oil has been reported to have therapeutic effects on chloasms, mellanosis, keratoderma, xeroderma, recurrent dermatitis, and other skin conditions when

Table 13.1. Sterol content (mg/kg) in seeds and pulp of two subspecies of *H. rhamnoides* (from Yang et al. 2001).

Sterol	Seed		Pulp	
	subsp. *sinensis*	subsp. *rhamnoides*	subsp. *sinensis*	subsp. *rhamnoides*
Campesterol	33 ± 4	40 ± 8	4.8 ± 1.3	3.8 ± 0.5
Stigmastadievol	–	–	5.5 ± 1.3	3.3 ± 0.5
Sitosterol	934 ± 105	931 ± 179	218 ± 74	198 ± 24
Stigmastanol	46 ± 11	50 ± 24	2.3 ± 1.5	1.3 ± 0.5
Isofucosterol + Obtusifoliol	244 ± 56	201 ± 55	–	–
Isofucosterol + Obtusifoliol + Stigmasta-8-en-3β-ol	–	–	15.3 ± 4.3	18.8 ± 8.5
Unidentified sterol	33 ± 19	5 ± 2	–	7.0 ± 2.8
Not identified + stigmasta-5,24(25)-dien-3β-ol	22 ± 4	17 ± 7	–	–
Not identified + stigmasta-5,24(25)-dien-3β-ol + stigmasta-(8,24)-dien-3β-ol	–	–	3.8 ± 1.7	5.0 ± 2.8
Stigmast-7-en-3β-ol + cycloartenol	25 ± 1	19 ± 13	8.5 ± 1.3	5.3 ± 1.5
4α, 14α-dimethyl-9β, 19-cycloergost-24(24')-en-3β-ol	12 ± 4	17 ± 9	10.3 ± 6.1	6.5 ± 3.1
Stigmasta-7, 24(24')-dien-3β-ol	26 ± 7	12 ± 3	7.6 ± 2.1	5.5 ± 1.7
Unidentified Sterol	–	–	5.0 ± 1.2	4.8 ± 3.1
4α, 14α, 24'-trimethylergosta-8, 24(24')-dien-3β-ol	7 ± 2	7 ± 3	4.8 ± 2.5	6.5 ± 1.7
24-methyl-5α-cycloart-24(24')-en-3β-ol(24-methylenesycloartanol)	21 ± 13	29 ± 19	9.3 ± 6.8	6.5 ± 2.4
Unidentified Sterol	4 ± 2	4 ± 2	–	3.3 ± 1.2
Citrostadienol	42 ± 17	39 ± 14	7.5 ± 2.0	5.8 ± 1.0
Unidentified sterol	–	–	6.5 ± 1.7	4.5 ± 1.7
Total	1441 ± 52	1369 ± 281	309 ± 67.6	279 ± 35.4

± values are standard deviations; – none reported.

tested in 350 people in China (He et al. 1989). It is also used in cosmetics as a natural plasticizer and emulsifier, in skin-regenerating compositions, and as natural UV filters.

In a double-blind, placebo-controlled study of 49 atopic dermatitis patients who received 5 g of sea buckthorn seed oil, pulp oil, or paraffin oil daily for 4 months, Yang et al. (1999) found significant improvement in groups receiving pulp oil and paraffin oil. Subjects consuming sea buckthorn seed oil showed no improvement in dermatitis (Yang et al. 2000). Subjects receiving pulp oil showed a small but significant increase in HDL cholesterol level (Yang et al. 1999). In a following, randomized, placebo-controlled, double-blind study, 5 g of sea buckthorn seed oil, pulp oil, or paraffin oil was administered to 16 patients with atopic dermatitis for 4 months (Yang et al. 2000). Skin fatty acids were measured. The group supplemented with seed oil showed slightly increased proportions of decosapentaenoic acid (22:5n-3) and decreased proportions of palmitic acid in skin glycerophospholipids. The high levels of linoleic and palmitoleic acids in seed and pulp oil, respectively, did not influence the fatty acid levels in skin glycerophospholipids (Yang et al. 2000) on short-term dietary modification. Sea buckthorn oil can also be used as cleanser (Zou 1997) and a cosmetic agent for the lips (Gercikovs and Zoludeva 1996).

Sea buckthorn fruit fractions, especially the oil, have been used in traditional medicine for treating cardiovascular disease both in animal experiments and clinical investigations (Xu and Chen 1991; Jiang et al. 1993; Li and Wang 1994). Sea buckthorn seed oil has been administered to both humans and animals with hyperlipidemia, resulting in a reduction of total cholesterol and triacylglycerol in plasma and an increase in the high density lipoproteins (Jiang et al. 1993). Supplementation with sea buckthorn oil has been reported to normalize the plasma lipid levels of hyperlipidemic subjects (Jiang et al. 1993). Recently, Johanssen et al. (2000) investigated the effects of C_2 extracted sea buckthorn fruit oil and combined sea buckthorn pulp and seed oil on some risk factors of cardiovascular disease. Eleven healthy normolipidemic men consuming 5 g of sea buckthorn oil per day for 4 weeks showed a clear decrease in the rate of adenosine-5-diphosphate-induced platelet aggregation along with reduced maximum platelet aggregation. This suggests a beneficial effect on blood-clotting and by extension, a beneficial effect on the cardiovascular risk factors. Blended (seed and pulp) sea buckthorn oil is a rich source of C_{16} and C_{18} unsaturated fatty acids, as well as vitamin E (tocopherols), carotenoids, and plant sterols (Beveridge et al. 1999). Sitosterol has been reported to inhibit platelet aggregation (Zhao et al. 1990) so that a relationship may exist between composition, nutritional value, and medicinal value, since sitosterol is present in sea buckthorn oil in large amounts. Phytosterols in sea buckthorn are listed in Table 13.1 for both seeds and pulp. These compounds are capable of reducing the risk factors of cardiovascular disease as noted above. The concentration of phytosterols is much higher in the seed than in the pulp, although the relative quantities of pulp and seed in the fruit should be considered when calculating the total dietary load contributed by sea buckthorn oils. In both fruit parts the major constituent is sitosterol which is capable of reducing platelet aggregation (Johansson et al. 2000) and in the seed isofucosterol and/or obtusifoliol provide additional significant contributions.

Sea buckthorn oil possees anti-tumor properties, since the seed oil is known to retard tumour growth by 30–50% (Wang et al. 1989). Seed oil (1.59 g/kg body weight) injected intraperitoneally significantly (30%) inhibited the growth rate of transplanted melanoma (B16) and sarcoma (S180) tumors in mice, without any influence on thymus or spleen weights (Zhang, W. et al. 1989a). It also increases the phygocytic rate and the index of macrophages significantly. The anti-tumor activity of oil from fruit residues (pulp) of sea buckthorn has been demonstrated by the significant increase in survival of mice bearing Ehrlich ascites carcinoma (Yang et al. 1989). The survival time and life span increase in a dose-dependant manner when the oil at 125–500 mg/kg was administered intraperi-

Table 13.2. Fatty acid composition of oils from seeds, fruits, and pulp of *H. rhamnoides* fruits of different origins (weight %) (from Yang and Kallio 2001).

Fatty acid	Seed oil		Fruit oil		Pulp oil	
	subsp. sinensis	subsp. rhamnoides	subsp. sinensis	subsp. rhamnoides	subsp. sinensis	subsp. rhamnoides
0.6666667	8.7 ± 0.7	7.4 ± 0.4	22.9 ± 4.6	23.6 ± 3.4	26.7 ± 4.5	27.8 ± 3.4
16:1n–7	–	–	21.5 ± 6.5	2.60 ± 3.3	27.2 ± 6.4	32.8 ± 3.2
0.75	–	–	1.5 ± 0.3	1.2 ± 0.1	1.3 ± 0.3	0.8 ± 0.2
18:1n–9	19.4 ± 3.7	17.1 ± 2.0	17.6 ± 3.5	17.2 ± 2.0	17.1 ± 3.8	17.8 ± 2.3
18:1n–7	2.2 ± 0.2	2.8 ± 0.7	6.7 ± 1.1	7.8 ± 0.6	8.1 ± 1.2	9.1 ± 0.8
18:2n–6	40.9 ± 1.7	39.1 ± 2.1	18.6 ± 6.8	15.8 ± 3.4	12.7 ± 6.9	9.0 ± 3.1
18:3n–3	26.6 ± 4.7	30.6 ± 3.0	11.2 ± 3.7	8.8 ± 2.1	7.1 ± 3.4	3.2 ± 1.6

± are standard deviations; – none reported.

toneally. Fractions extracted from the oil have no significant cytotoxic activity (50%) on human leukemia cell strain K562 at a concentration of 25 mg/mL (Yang et al. 1989). The protective effect of sea buckthorn seed oil toward cervical cancer was thought due to the presence of vitamins A and E and β-carotene (Wu et al. 1989)

The maximum dose for rats of sea buckthorn seed oil, while avoiding undesirable effects, is 9.5 g/kg body weight (i.e., safe for long-term use). Adverse or toxic effects have not been observed with sea buckthorn oil even when stored at 18°C for 2 years, and at doses 10–20 times greater than the maximum therapeutic dose. In fact, the animals gained weight. No toxic effects were observed in acute (10 g/kg body wt) as well as chronic (0.5 g/kg for a month) consumption tests (Mironov et al. 1989). When administered to mice, rats, cats, and dogs at 20–30 g/kg body weight by single or repeated injections over 2 months, sea buckthorn oil did not influence blood chemistry (number of erythrocytes, haemoglobin, etc.) or the pathomorphology of animal organs (liver, stomach, kidney, spleen, lymphatic nodes, thyroid glands, and adrenals) (Rachimov et al. 1989).

Sea buckthorn oil is a good source of α-linolenic acid (18:3n-3) and other unsaturated fatty acids (Table 13.2). The potential of α- and γ-linolenic acids to assist in the treatment of such conditions as dermatitis, rheumatoid arthritis and platelet aggregation in clinical applications is discussed by Barre (2001) for evening primrose, borage, black currant, and fungal oils, all high in α- and γ-linolenic acids. It was very difficult to present clear, convincing evidence for the efficacy of these oils in preventing human disease. In the light of repeated observations of the inhibition of platelet aggregation with sea buckthorn oil (Zhao et al. 1990; Johansson et al. 2000), it seems likely that this activity is not associated with these fatty acids. However, the oil has a unique composition of fatty acids, including a spectrum of mono-unsaturated acids and is a rich source of two essential fatty acids, linoleic and α-linolenic and both the seed and pulp oils are rich in oleic (18:1n-9) acid. There is increasing interest in the physiological role of mono-unsaturated acids relative to the modification of vascular endothelial surfaces and the role this may play in the reduction of thrombosis (Perez-Jimenez et al. 1999; Turpein et al. 1998), treat psoriasis (Hodutu 1999), and peptic ulcer (Abidov et al. 2000). Endothelial cells produce a large number of substances involved in adhesion and transendothelial migration of circulating leucocytes into the vascular cell wall, in addition to coagulation and fibrinolysis. All of these factors are involved in atherosclerotic development.

Chapter 14. Sea Buckthorn Ecology

Thomas S.C. Li[1] and E. Small[2]

[1] *Agriculture and Agri-Food Canada, Pacific Agri-Food Research Centre, Summerland, British Columbia V0H 1Z0*
[2] *Agriculture and Agri-Food Canada, Eastern Cereal and Oilseed Research Centre, Ottawa, Ontario K1A 0C6*

Sea buckthorn is very valuable for promoting wildlife, especially in its native range. Many animals use it for food and shelter. In the Canadian prairies, the shrubs provide valuable habitat for the native sharpedtailed grouse. Various birds have been shown to be effective at distributing the seed of sea buckthorn. It was reported that the germination rate of sea buckthorn seeds is six times greater when it has passed through a birds gut than it has not (Gillham 1987). While it is useful for promoting ecosystem welfare, there is concern about its invasiveness (Baker 1996) mainly due to the suckering characteristics after the trees are established (see Chapter 2). Sea buckthorn bush will increase its area at the rate of approximately 6% per year (Ranwell 1979). Over the years, the species has spread over almost the entire Eurasian continent from Lake Baikal and central China to the Himalayas, where it occurs in sizeable stands on flood-plains about 500 m above sea level (Ridley 1930; Rousi 1971). In Canada, sea buckthorn is not a native species, there are a few habitats created specially for wildlife, such as birds and small animals, and its spread only occurred within the habitat (Li, personal observation).

There are a few ways to control natural spreading of sea buckthorn. In orchards, herbicides application within the row, regular machine mowing and ploughing between the rows to prevent sucker development, and selection of cultivars that have less or no suckers produced. In the natural habitat, maintain zero *Hippophae* population of any sites currently free from *Hippophae* by uprooting seedlings as they are found. Where *Hippophae* is well established, control so as to maintain habitat diversity (Ranwell 1972). Biological controls seem to be appropriate in some regions. Larvae of the moth *Malocosini* have been known to defoliate sea buckthorn, and rabbit grazing has been observed to stunt the growth of suckers, although neither cause serious damage to the tree (Robinson 1972). It was reported that cattle and sheep may moderately control *Hippophae* spreading (Anonymous 1992), however, in Mongolia, sheep can not, or will not, go into areas of sea buckthorn habitat with very high density and only graze the edge of the habitat (Li, personal observation).

Sea buckthorn is a dioecious species, with seeds formed after female flowers are pollinated by wind-dispersed pollen. The resulting outcrossing produces considerable heterozygosity in regions with more than one species or subspecies (see Chapters 2 and 3). This may be the reason that there is extensive genetic variation in *H. rhamnoides* at subspecies, populational, and individual levels (Yao and Tigerstedt 1995). It is difficult to evaluate the extent to which pollen dispersal by wind brings about gene flow between isolated populations of *Hippophae*. However, strong winds may transport pollen over great distances and the resulting hybridization might tend to blur the boundaries between populations and taxa (Rousi 1971). Once in a given location, certainly sea buckthorn species, such as subsp. *sinensis*, spreads extensively by means of its extensive sucker system. The seed propagation of the species complement this localized vegetative system by adaptation to long-distance dissemination by fruit-eating birds to distant regions. This phenomenon obviously has had an effect on the distribution and variation of the taxa (Rousi 1971). In some regions, sea buckthorn seeds may be dispersed by wind, as reported by Müller (1995), who observed that birds did not eat sea buckthorn fruits in Graubunden, Switzerland. The distribution and variation of the taxa of *Hippophae* seem to have resulted in part from long-distance dispersal (Rousi 1971), but probably also reflect a geographical pattern of survival following the last glaciation. Sea buckthorn fruits persist on the fruiting branch over the winter until late spring as an achene. There were no reports of the viability of seeds remaining in dried fruits. Seeds can be seeded in the fall and seedlings start to emerge in the spring (see Chapter 6).

Sea buckthorn is a colonizer of open habitats. The European plants are typically found on slopes, riverbanks, and seashores. Over most of their natural distribution, the plants receive 30–40 cm of precipitation annually. The three European subspecies of *H. rhamnoides* are widespread, generally in low altitude areas under 2000 m. In contrast, most Asian representatives of the genus are mountain adapted and rarely found below 3000 m. Observations and surveys show that many birds and animals utilize sea buckthorn for food and shelter (Ma and Sun 1986; Salo 1989). In the Canadian prairies sea buckthorn is especially valuable habitat for the sharp-tailed grouse, Hungarian partridge, and pheasant (Schroeder 1995).

Sea buckthorn has a highly efficient symbiotic relationship with a bacterium of the genus *Frankia*, which belongs to a primitive class of bacteria, the Actinomycetes. This relationship, which is similar to that of the pea family, Fabaceae, with the relatively advanced bacteria of the genus *Rhizobium*, allows atmospheric nitrogen to be converted to a form that can be used for nutrition (Akkermans et al.1983; Dobritisa and Novik 1992). Only about two dozen other plant genera have developed such nitrogen-fixing relationships with the long, thread-like Actinomycetes (see Chapter 2). The resulting improved root growth enhances the entire soil ecosystem: there is more organic matter, more oxygen, and more soil organisms, which amounts to more soil biodiversity.

Sea buckthorn has no specific requirement of soil conditions, it can survive in marginal land (see Chapter 5) and is useful in reclaiming and conserving soil, especially on fragile slopes, due particularly to its extensive root system (Li and Schroeder 1996). It was reported that *Hippophae* was able to tolerate a sodium chloride solution (32 g/L) periodically sprayed on to the plants, or added to the soil, through two successive winters (Thompson and Rutter 1986), and seed germination was completely prevented by salinity greater than 0.05% (Pearson and Roger 1962). Because it is resistant to drought and tolerates soil salinity and low temperatures down to –43°C (see Chapter 2), it is suitable for many situations that are simply too demanding for most plants. Riverbanks, lakeshores, steep slopes, and other susceptible terrain can benefit from the establishment of sea buckthorn (Li and Schroeder 1996). Windbreaks made up of sea buckthorn are effective at preventing wind erosion in open areas. Not only does sea buckthorn prevent the loss of soil, but it also improves degraded soils due to its nitrogen-fixing capabilities (Schroeder and Yao 1995). Roots of sea buckthorn also are able to transform insoluble organic and mineral matter in the soil into more soluble states (Lu 1992). Thus, there is reduced need to add fertilizers, which results in less input costs as well as fewer ecological problems (see Chapter 2).

Ecosystems in the mountains of the developing countries of Asia, particularly in the Hindu Kush-Himalayan (HKH) Region, have been facing ecological imbalances and environmental degradation. Corrective efforts have been made in the past by various agencies, but the problems are so complex, critical, and widespread that the situation has deteriorated. The International Centre for Integrated Mountain Development (ICIMOD) was established in 1983 in Nepal to help promote the development of an economically and environmentally sound mountain ecosystem and to improve the living standards of the mountain populations of the HKH Region. In its efforts, ICIMOD has identified sea buckthorn as an extremely useful and multi-purpose plant capable of reducing the problems considerably. Over the years, it has been shown that sea buckthorn has contributed significantly towards the economic and environmental improvement in the HKH region.

Sea buckthorn has been useful in lessening pollution resulting from erosion of contaminated mine waste, since it can be used to re-vegetate a variety of mine spoils (Small et al. 2002). Because sea buckthorn is naturally resistant to pests, there is limited need for pesticides that are potentially damaging to the environment. In parts of North America, it has been planted as cover along highways where de-icing salt prevents growth of many other woody plants. Thus, sea buckthorn helps to prevent and reduce erosion and dissemination of pollutants from roadsides (Small et al. 2002).

China is confronted with grave problems of soil erosion, with 2 million sq km of eroding land. China has estimated its annual soil losses to be around 5 billion tons and the annual loss of cultivated land about 70 000 ha. The most promising tool to control land degradation in China is re-vegetation, and sea buckthorn is one of the species successfully used on a large scale. In northern China, it is helping to control desertification, conserve land and water resources, and integrate economic exploitation with ecological rehabilitation. Over 1 million ha of sea buckthorn have been planted in China, most of it for soil and water conservation.

A living windbreak is a linear arrangement of plants, primarily trees and shrubs, established to reduce the damaging effects of strong winds and soil erosion. It also helps to protect crops, reduce snow drifting, and create wildlife habitat. Plants that serve as windbreaks must be resistant to the drying effects and physical injuries caused by wind, and sea buckthorn is well suited to this task. It has been grown on the Canadian prairies since the late 1970s as part of farmstead shelterbelts. Over 1 million seedlings of sea buckthorn have been distributed through the shelterbelt program of Canada's Prairie Farm Rehabilitation Administration since 1982.

Sea buckthorn, despite being a common and widespread species, is deserving of conservation measures. Its distribution pattern has been described as highly fragmented and isolated populations are often genetically distinct. A recent analysis of sea buckthorn genetic resources noted that the high nutrient and medicinal values of the fruit have led to uncontrolled exploitation and even destruction of sea buckthorn resources in some parts of its natural distribution. Thus, protection and preservation of the valuable germplasm of the genus has become urgent. In Hungary, wild sea buckthorn is rarely observed, and the plant is protected as an endangered species. Such protection needs to be extended, especially to the Asian sea buckthorns, which occupy small distinct ranges. As a newly deomesticated fruit crop, sea buckthorn requires considerable research and development to be sustainable, and its wild, worldwide population constitute an invaluable genetic resource for selecting superior plants.

References

Abartene, D.Y., and Malakhovskis, A.I. 1975. Combined action of a cytostatic preparation and sea buckthorn oil on biochemical indices, part 2 hisphen. Liet. TSR Mokslu Akad. Darb. Ser. C Biol. Mokslai **1**: 167–171.

Abidov, M.T., Zhilov, V.Kh., Ovchinnikov, A.A., and Khokhlov, A.P. 2000. Peptic ulcer treatment: Galavit combined with tocopherol and kaolin or almagel in rose or sea buckthorn oil. Russian Patent (RU 2160104).

Affeldt, H.A., Marshall, D.E., and Brown, G.K. 1988. Relative dynamic displacements within a trunk shaker clamp. Trans. ASAE (Am. Soc. Agric. Eng.) **31**: 331–336.

Agafonova, V.N., and Borodachev, M.N. 1986. Features of the accumulation of biologically active compounds in the fruits of sea buckthorn in Moscow Provinces. Sostoyanie Perspekt. Razvit. Kul't. Oblepiikhi Necharnozenmoi Zone RSFER: 53–59.

Agrawal, A., Khuntia, B.B., and Dubey, G.P. 2001a. Sea buckthorn modulates the mental health status and its effect on immunoscenescence among the elderly–a double blind placebo control study. In Proceedings of an International Workshop on Sea Buckthorn. A Resource for Health and Environment in Twenty First Century, February 18–21, 2001, New Delhi, India. pp. 213–220.

Agrawal, A., Tewari, B.S., and Dubey, G.P. 2001b. Role of seabuckthorn (Hippophae rhamnoides) in the management of age related memory disorder. In Proceedings of an International Workshop on Sea Buckthorn. A Resource for Health and Environment in Twenty First Century, February 18–21, 2001, New Delhi, India. pp. 225–230.

Agrawal, A., Dixit, S.P., Mishra, A., and Dubey, G.P. 2001c. Determination of safety and efficacy profile of seabuckthorn (Hippophae rhamnoides L.). In Proceedings of an International Workshop on Sea Buckthorn. A Resource for Health and Environment in Twenty First Century, February 18–21, 2001, New Delhi, India. pp. 235–238.

Akkermans, A.D.L., Roelofsen, W., Blom, J., Huss-danell, K., and Harkink, R. 1983. Utilization of carbon and nitrogen compounds by Frankia in synthetic media and in root nodules of Alnus glutinosa, Hippophae rhamnoides, and Datisca cannabina. Can. J. Bot. **61**: 2793–2800.

Albrecht, H.J. 1993. New demands faced by buckthorn breeding. In Cultivation and utilization of wild fruit crops. Bernhard Thalacker Verlag GmbH & Co. pp. 57–60. [In German]

Albrecht, H.J., Gerber, J., Koch, H.J., and Wolf, D. 1984. Experience in growing sea buckthorn. Gartenbau **31**: 242–244. (from Hortic. Abst. 54: 9033).

Andreeva, I.N., Fedorova, E.E., Il'yasova, V.B., and Tibilov, A.A. 1982. Ultrastructure of nitrogen-fixing and wintering nodules in one-year seedlings of sea buckthorn and oleaster, Hippophae rhamnoides, Elaeagnus argentea. Sov. Plant Physiol. (USA) **29**: 109–116.

Anonymous. 1992. Sand dune grazing seminar notes. Peterborough: English Nature.

Anonymous. 1994. Weed control in shelterbelts. Prairie Farm Rehabilitation Administration, Agric. Agri-Food Can. Publ. T.N. Circ. 6.

Anonymous. 2000. Phytosterol esters (plant sterol and stanol esters). Food Sci. Technol. Today **14**: 154–158.

Anonymous. 2001a. Draco Natural Products launches sea buckthorn. HAPPI (Household & Personal Products Industry) **38**(11): 114.

Anonymous. 2001b. Tagra microencapsulates vitamins & natural oils. HAPPI (Household & Personal Products Industry) **38**(11): 145.

Araujo, P.E. 1977. Role of citrus fruit in human nutrition. In Citrus Science and Technology,

Vol. 1. Nutrition, Anatomy, Chemical Composition and Bioregulation. *Edited by* S. Nagy, P. E. Shaw, and M.K. Veldhuis. AVI Publ. Co. Inc., Westport, CT. p. 13.

Arkhipova, T.N., Pipunyrov, S.V., and Popov, V.A. 1995. Production of sea buckthorn oil by extraction of dried milled sea buckthorn fruit with dichloro-difluoro-methane, followed by distillation. Russian Patent 207239.

Avdeev, V.I. 1976. Propagation of *Hippophae rhamnoides* by softwood cuttings under mist. Kratkie Tezisy Dokl 2-i Vses. Konf. Molodykh Uchenykh po Sadovodstvu (1976) 72–74. (from Hortic. Abst. **48**: 318).

Avdeev, V.I. 1984. Propagation of fruit crops by cuttings in upper Tadjikistan. Intensivnye Sposoby Vyrashchivaniya Posadochnogo Materiala Sadovykh Kul'tur (1984): 51–56. (from Hortic. Abst. **55**: 40).

Bailey, L.H. and E.Z. Bailey. 1978. Hortus Third, A concise dictionary of plants cultivated in the United States and Canada. MacMillan Pub. Co. Inc. 1290 p.

Baker, R.M. 1996. The future of the invasion shrub, sea buckthorn (*Hippophae rhamnoides*), on the west coast of Britain. Asp. Appl. Biol. **44**: 461–468.

Balabushka, V.K. 1990. Results of the trials of growth regulators on summer cuttings of introduced woody plants. Byull. Gl. Bot. Sada **156**: 65–67. (from Hortic. Abst. **61**: 5083).

Balint, K., Terpo, A., and Zsoldos, L. 1989. Sea buckthorn as suitable plant for reclamation of red mud impoundments in Hungary. *In* Proceedings of the First International Symposium on Sea Buckthorn, October 19–23, 1989, Xi'an, China. pp. 268–274.

Barre, D.E. 2001. Potential of evening primrose, borage, black currant, and fungal oils in human health. Ann Nutr. Metab. **45**: 47–57.

Bartish, I. V., Jeppsson, N., Nybom, H., and Swenson, U. 2002. Phylogeny of Hippophae (Elaeagnaceae) inferred from parsimony analysis of chloroplast DNA and morphology. Syst. Bot. **27**: 41–54.

Bekaso, L.S. 1992. Production of oil from sea buckthorn berries by milling, extraction with vegetable oil and separation of juice after settling. Russian Patent 2032720.

Beldean, E.C., and Leahu, I. 1985. *Hippophae rhamnoides*–a valuable fruit-producing pioneer species. Forstarchiv 56: 249–253.

Bencharov, J.V. 1999. Cosmetic cream. Russian Patent 2134570.

Berezhnaya, G.A., Ozerinina, O.V., Yeliseev, I.P., Tsydendambaev, V.D., and Vereshchagin, A.G. 1993. Developmental changes in the absolute content and fatty acid composition of acyl lipids of sea buckthorn fruits. Plant Physiol. Biochem. **31**: 323–332.

Berezhnaya, G.A., Eliseev, I.D., Tsydendambaev, V.D., and Vereshchagin, A.G. 1989. The effect of using foreign pollen on the content and composition of acyl lipids in the sea buckthorn (*Hippophae rhamnoides* L.) fruits. *In* Proceedings of the First International Symposium on Sea Buckthorn, October 19–23, 1989, Xi'an, China. pp. 163–166.

Bernath, J., and Foldesi, D. 1992. Sea buckthorn (*Hippophae rhamnoides* L.): A promising new medicinal and food crop. J. Herbs Spices Med. Plants **1**: 27–35.

Beveridge, T., 1997. Juice extraction from apples and other fruits and vegetables. Crit. Rev. Food Sci. Nutr. **37**: 449–469.

Beveridge, T. 2000. Large Scale Centrifugation. *In* Encyclopedia of Separation Science. *Edited by* I.D. Wilson, E.R. Adlard, M. Cooke, and C.F. Poole. Academic Press, London.

Beveridge, T. 2002. Opalescent and Cloudy Fruit Juices: Formation and Particle Stability. Crit. Rev. Food Sci. Nutr. **42**: 317–337.

Beveridge, T., Li, T.S.C., Oomah, B.D., and Smith, A. 1999. Sea Buckthorn Products: manufacture and Composition. J. Agric. Food Chem. **47**: 3480–3488.

Beveridge, T., Harrison, J.E., and Drover, J. 2002. Processing effeects on the composition of sea buckthorn juice from *Hippophae rhaminoides* L. cv. Indian Summer. J. Agric. Food Chem. **50**: 113–116.

Bhagat, R.M., Kashyap, N.P., and Singh, V. 2001. Insect-pests associated with sea buckthorn (*Hippophae rhamnoides* L.) in Lahaul Valley, dry temperate Himalayas. *In* Proceedings of the International Workshop on Seabuckthorn, Feb. 18–21, 2001, New Dehli, India. pp. 133–135.

Biswas, M.R., and Biswas, A.K. 1980. In desertification, control the deserts and create pastures. Environ. Sci. Appl. **12**: 145–162.

Bock, W., Felkenheuer, W., Dongowski, G., Kroll, J., Schveider, C., Baars, H., and Sievert, B. 1990. Method for enhanced processing of raw juice from sea buckthorn berries. GDR Patent DD 275 775 A3.

Bond, G. 1983. Taxonomy and distribution of non-legume nitrogen-fixing systems. *In* Biological nitrogen fixation in forest ecosystems: foundations and applications. *Edited by* J.C. Gordon and C.I. Wheeler. Martinus Nijhoff, Dr. W. Junk Publishers, The Hague. pp. 55–87.

Botenkov, V.P., and Kuchukov, N.M. 1984. Device for collecting fruits of *Hippophae rhamnoides*. Lesn. Khoz. **2**: 46–47.

Bounous, G., and Zanini, E. 1988. The variability of some components and biometric characteristics of the fruits of six tree and shrub species. *In* Lampone, mirtillo ed altri piccoli frutti Atti, convegno, trento, 1987. Ministero agricoltura e foreste, Rome, 1988. pp. 189–197. (from Hortic. Abst. **60**: 4153).

Buglova, T.L. 1978. Fruit size in some ecological and geographical groups of *Hippophae rhamnoides* and the inheritance of large-fruitedness in the hybrid progeny. Nauchn. T. Omsk. S-kh. Inst. **175**: 26–29. (from Plant Breed. Abst. **50**: 4357).

Buglova, T.L. 1981. Effect of male plants of *Hippophae rhamnoides* on the yield of a plantation. Lesn. Khoz. **11**: 39–40.

Buhatel, T., Vesa, S., and Morar, R. 1991. Data on the action of sea buckthorn oil extract in the cicatrization of wounds in animals. Bul. Inst. Agron. Cluj-Napoca Ser. Zooteh. Med. Vet. **45**: 129–133.

Bump, V.L. 1989. Apple Pressing and Juice Extraction. *In* Processed Apple Products. *Edited by* D.L. Downing. Avi Publishing Co., Van Nostrand Reinhold, New York and London. pp. 53.

Burdasov, V.M., and Sviridenko, E.I. 1988. Production of regenerates of sea buckthorn from apical meristems in in-vitro culture. Sib. Vestn. Sel'skokhozyaistvennoi Nauk. **3**: 106–110.

Business Wire. 1999. Studies show anti-ulcer and anti-cancer activity in sea buckthorn seed oil. (October 1999)

Centenaro, G., Capietti, G.P., Pizzocaro, F., and Marchesini, A. 1977. The berry of the sea buckthorn Hippophae rhamnoides as a source of vitamin C. Atti Soc. Ital. Sci. Nat. Mus. Civ. Stor. Nat. Milano **118**: 371–378.

Chai, Q. 1989. The experimental studies on the cardiovascular pharmacology of sea buckthorn extract from *Hippophae rhamnoides* L. *In* Proceedings of the First International Symposium on Sea Buckthorn, October 19–23, 1989, Xi'an, China. pp. 392–397.

Chen, J.H. 1991. Effect of the immunomodulating agents (BCG) and the juice of HRL on the activity of splenic NK cells and LAK cells from tumor bearing mice. Chin. J. Microbiol. Immunol. **11**: 105–108.

Chen, C., Lui, B., and Yu, Y. 1995. Studies on the pigment of sea buckthorn. Hippophae **7**: 34–40.

Chen, Y. 1989. The volatile compounds in the sea buckthorn oil and its saponification products. *In* Proceedings of the First International Symposium on Sea Buckthorn, October 19–23, 1989, Xi'an, China. p.75–80.

Chen, T. 1988. Preliminary research on the biochemical components of sea buckthorn oil from Gansu. Seabuckthorn **1**: 35–38.

Chen, T.G., Ni, M.K., Li, R., Ji, F., Chen, T., Ni, M.K., Li, R., and Ji, F. 1991. Investigation of the biological properties of Central Asian Sea Buckthorn growing in the province of Kansu, China. Chem. Nat. Compd. **27**: 119–121.

Chen, Y.D., Jiang, Z.R., Qin, W.L., Ni, M.N., Li, X.L., and He, Y.R. 1990. Research on the chemical composition and characteristics of sea buckthorn berry and its oil. Chem. Ind. For. Prod. **10**: 163–175. [In Chinese]

Cheng, T., Li, T., Duan, Z., Cao, Z., Ma Z., and Zhang, P. 1990. Acute toxicity of flesh oil of *Hippophae rhamnoides* L. and its protection against experimental hepatic injury. Chung Kuo Chung Yao Tsa Chih 15: 45–47.

Chistjakov, A.G. 1998. Face cream "Talita". Russian Patent 2123320.

Cireasa, V. 1986. *Hippophae rhamnoides* L. extension on Rufeni Hill, Iasi district. Lucr. Stiint. Inst. Agron. "Ion Ionescu de la Brad", Horticult. **30**: 75–77. (from Hortic. Abst. **58**: 6535).

Crapatureanu, S., Rosca, G., Neamtu, G., Socaciu, C., Britton, G., and Pfander, H. 1996. Carotenoids from sea buckthorn fruit determined by thin layer chromatography and mass spectrometry. Rev. Roum. Biochem. **33**: 167–174.

Dai, Y.R., Gao, C.M., Tian, Q.L., and Yin, Y. 1987. Effect of extracts of some medicinal plants on superoxide dismutase activity in mice. Planta Med. **53**: 309–310.

Davidson, C.D., Enns, R.J., and Gobin, S. 1994. Landscape plants at Morden Arboretum. Agriculture & Agri-Food Canada, Morden, Manitoba, Canada. 163 p.

Degering, F. (*Editor*). 1963. Organic Chemistry. 6th ed. Barnerousis and Noble, New York. p. 80.

Degtyareva, I.I., Toteva, E.T., Litinskaya, E.V., Matvienko, A.V., Yurzhenko, N.H., Leonov, L.N., Khomenko, E.V., and Nevstruev, V.P. 1991. Lipid peroxidation level and vitamin E concentrations in the treatments of ulcer patients. Klin. Med. (Mosc.) **69**: 38–42.

Demenko, V.I., Mikityuk, O.D., and Levinskii, M.B. 1986. Abscisic acid, ethylene, growth and fruit drop in sea buckthorn. Fiziol. Rast. (Mosc.) **33**: 188–194. (from Hortic. Abst. **56**: 7655).

Demenko, V.I., and Korzinnikov, Y.S. 1990. Effect of surface-activity compounds and copper ions on ethylene producers inducing fruit abscission in the sea buckthorn. Fiziol. Rast. (Mosc.) **37**: 596–601.

Detsina, A.N., and Selivanov, B.A. 1998. Winter cosmetic cream for skin protection. Russian Patent 2120272.

Dobritsa, S.V., and Novik, S.N. 1992. Feedback regulation of nodule formation in *Hippophae rhamnoides*. Plant Soil **144**: 45–50.

Egyed-Balint, K., and Terpo, A. 1983. Effect of red mud on growth and element accumulation in some plant species. Kerteszeti Egyetem Kozlemenyei **47**: 127–136.

Eliseev, I.P. 1976. Biologically active substances in the fruits of *Hippophae rhamnoides* in Central Asia and the Caucasus. Biol. aktiv. veshchestva plodov i yagod. (1976) 161-163. (from Plant Breed. Abst. **48**: 5917).

Eliseev, I.P., and Mishulina, I.A. 1972. The effect of the minor elements on the increase in sea buckthorn (*Hippophae rhamnoides*) seed germination. Tr. Gor'k. S-Kh Inst. **38**: 110–111.

Eliseev, I.P., and Fefelov,V.A. 1977. Material for studying *Hippophae rhamnoides* in Kabardino-Balkaria. Tr. Gor'k. S-Kh Inst. **105**: 3–7.

Faber, H. 1959. Weed control in woody ornamental nurseries and seedbeds. Mitt. Biol. Zent. Anst. Berl. **97**: 144–148.

Fayman, B.A. 1991. Treatment of operative wounds in ear, nose and throat with sea buckthorn oil. Seabuckthorn **4**: 7.

Fefelov, V.A., and Eliseev, I.P. 1986. Biology of germination and emergence of sea buckthorn seeds of different ecological and geographical origin. Biol. Khim. Introd. i sellekrsiya Oblopikhi **1986**: 110–115.

Flavex. 1992. CO_2- extracts. Prospectus of FLAVEX Naturextrakte, Rehlingen, Germany.

Franke, W., and Müller, H. 1983*a*. Quantity and composition of fatty acids in the fat of the juicy fruit part and of the seed of fruits of *Hippophaë rhamnoides*. L. Angew. Bot. **57**: 77–83.

Franke, W., and Müller, H. 1983*b*. A contribution to the biology of useful plants. 2. Quantity and composition of fatty acids in the fat of the fruit flesh and seed of sea buckthorn. Angew. Bot. **57**: 77–83.

Frolov, V.V. 1999. Ointment for treating burns and infected injuries. Russian Patent 2129423.

Gaetke, R., and Triquart, E. 1992. Pruning machine for mechanized harvest of sea buckthorn. Gartenbau Mag. **1**: 57–58.

Gaetke, R., and Triquart, E. 1993. First results with an improved sea buckthorn harvesting technology. *In* Cultivation and utilization of wild fruit crops. Bernhard Thalacker Verlag GmbH & Co. pp. 37–41. [In German]

Gaetke, R., Schmidt, M., Triquart, E., and Wegert, F. 1991. Ernteverfahren sanddorn. Erwerbsobstbau **2**: 49–51.

Gakov, M.A. 1980. Prospects for the development of *Hippophae rhamnoides* in the Tuva, USSR. Lesn. Khoz. **2**: 51–52.

Gao, X., Ohlander, M., Jeppsson, N., Bjork, L., and Trajkovski, V. 2000. Changes in antioxidant effects and their relationships to phytonutrients in fruits of sea buckthorn (*Hippophae rhamnoides* L.) during maturation. J. Agric. Food Chem. **48**: 1485–1490.

Garanovich, I.M. 1984. Features of the vegetative propagation of *Hippophae rhamnoides* for introduction in Belorussia. Lesn. Khoz. **2**: 27–29.

Garanovich, I.M. 1995. Introduction and selection of *Hippophae* in Belarus. *In* Proceedings of the International Workshop on Sea Buckthorn, December 1995, Beijing, China.

Gatner, E.M.S., and Gardener, I.C. 1970. Observations on the fine structrure of the root nodule endophyte of *Hippophae rhamnoides* L. Arch. Mikrobiol. **70**: 183–196.

Gavrishin, N.S., Gavrishin, S.D., and Voronoskii, A.V. 1990. Production of sea buckthorn oil with increased carotenoid content by milling sea buckthorn fruit, separating juice from solids, extracting juice with vegetable oil and extracting solids with obtained extract. Russian Patent 2053255.

Gercikovs, L., and Zoludeva, I. 1996. Cosmetic agents for the lips. Patent in Latvia (LV 12065).

Gillham, M. E. (*Editor*). 1987. Sand Dunes. Heritage Coast Joint Management Advisory Committee publication. pp. 27–41.

Gladon, R.J., Hannapel, D.J., and Kolomiets, M.V. 1994. Small-fruit and tree-fruit research centers in Ukraine. HortScience **29**: 1214, 1393–1395.

Glazunova, E.M., Mukhtarova, E. Sh,. Zakharov, K.S., and Gachechiladze, N.D. 1994. New cycloartane terpenoid from the oil of sea buckthorn fruit. Chem. Nat. Compd. **30**: 271–272.

Glazunova, E.M., Gachechiladze, N.D., Bondar, V.V., Korzinnikov, Y.S., Potapova, I.M., and Gur'yanov, A.F. 1984. Biochemical fruit characteisitics of *Hippophae rhamnoides* L. growing in the Western Pamirs. Rast. Resur. **20**: 232–235. (from Hortic. Abst. **54**: 6135).

Goncharov, P.L. 1995. Sea buckthorn in Siberia: problems and solutions. *In* Proceedings of the International Workshop on Sea Buckthorn, December 1995, Beijing, China. pp. 97–105.

Goncharova, N.P., and Glushenkova, A.I. 1995*a*. Polar lipids of *Hippophae rhamnoides* leaves. Chem. Nat. Compd. **31**: 562–564.

Goncharova, N.P., and Glushenkova, A.I. 1995*b*. Epicuticular and intracellular lipids of *Hippophae rhamnoides* leaves. Chem. Nat. Compd. **31**: 665–671.

Goncharova, N.P., and Glushenkova, A.I. 1996. Lipids of the leaves of two forms of Central Asian sea buckthorn. Chem. Nat. Compd. **32**: 585–586.

Gontea, I., and Barduta, Z. 1974. The nutrient value of fruits from *Hippophae rhamnoides*. Igiena **23**: 13–20. (from Nutr. Abst. Rev. **45**: 8559).

Gorunzhina, S.I., Pavlov, S.S., and Sharygin, M.A. 1990. Production of sea buckthorn oil by mixing sea buckthorn fruit with preheated vegetable oil, milling in centrifugal field, separating intact seeds and extracting. Soviet Union Patent 1833409.

Gurevich, S.K. 1956. The application of sea buckthorn oil on ophthamology. Vestn. Ottamol. **2**: 30–33.

Hakkinen, S.H., Karenlampi, S.O., Mukkanen,

H.M., and Torronen, A.R. 2000. Influence of domestic processing and storage on flavonol contents in berries. J. Agric. Food Chem. **48**: 2960–2965.

Harju, K., and Ronkainen, P. 1984. Metals in Finnish liqueurs. Z. Lebensm.-Unters. Forsch. **178**: 393–396.

Harrison, J.E., and Beveridge, H.J.T. 2002. Sea buckthorn cv. Indian Summer: Fruit structure. Can J. Bot. **80**: 399–409.

He, X., Fang, Z., Liang, Z., and Jiang, L. 1989. Development and utilization of sea buckthorn fruit cosmetic. *In* Proceedings of the First International Symposium on Sea Buckthorn, October 19–23, 1989, Xi'an, China. pp. 320–321.

Heilscher, K., and Bat, S. 1990. Prerequisites for extracting sea buckthorn oils. Gartenbau **37**: 219–220.

Heilscher, K., and Lorber, S. 1996*a*. Process for the manufacture of oil-containing, haze stable semi-manufactured products from fruits with oily flesh, especially sea buckthorn berries and their use. FRG Patent DE 44 31 394 C1. (Food Sci. Technol. Abst. 6J261).

Heilscher, K., and Lorber, S. 1996*b*. Cold working process for obtaining clear juice, sediment and oil from sea buckthorn berries and their use. GFR Patent DE 44 31 394 C1.

Heinze, M., and Fiedler, H.J 1981. Experimental planting of potash waste dumps. 1. Communication: pot experiments with trees and shrubs under various water and nutrient conditions. Arch. Acker- Pflanzenbau Bodenkd. **25**: 315–322.

Hirrsalmi, H. 1993. The role of natural small fruits in Finnish plant breeding. Aquilo. Ser. Bot. **31**: 59–67.

Hodutu, M. 1999. Medicinal preparations containing catina oil and vitamins for treatment of psoriasis. Patent in Canada (CA 2209625).

Horvarth, J., Kelemen, J., Lassu, I., Marko, G., Mozsik, G., and Sarudi, I. 1994. Oil extraction of sea buckthorn berries for human consumption and cosmetics. Hungarian Patent 72703.

Hoy, E.C., and Su, X. 2001. Structured triacylgly urens. *In* Structured and Modified Lipids. *Edited by* F.D. Gunstone. Marcel Dekker Inc., New York. pp. 209–239.

Huang, Q. 1995. A review on *Hippophae* breeding in China. *In* Proceedings of the International Workshop on Sea Buckthorn, December 1995, Beijing, China.

Ivanicka J. 1988. Propagation of unusual fruit crops from softwood cuttings under mist. Ved. Pr. Vysk. Ust. Ovocnych Okrasnych Drev. Bojniciach **7**: 163–170. (from Hortic. Abst. **59**: 3687).

Jeppsson, N., Gao, X.Q., and Gao, X.Q. 2000. Changes in the contents of kaempherol, quercetin and L-ascorbic acid in sea buckthorn berries during maturation. Agri. Food Sci. Finl. **9**: 17–22.

Jiang, Y.D., Zhou, Y.C., Bi, C.F., Li, J.M., Yang, J.X., Yu, Z.D., Hu, Z.Y., and Zhao, S.X. 1993. Clinical investigation of effects of sea buckthorn seed oil on hyperlipidemia. Hippophae **6**: 23–24.

Jiang, Z., Qian, D., and Chou, S. 1989. An experimental study of *Hippophae rhamnoides* seed oil against gastric ulcer. *In* Proceedings of the First International Symposium on Sea Buckthorn, October 19–23, 1989, Xi'an, China. pp. 401–402.

Jin, Y. 1989. Super-oxide dismutase firm fruit foliage of common sea buckthorn *Hippophae rhamnoides* L. *In* Proceedings of the First International Symposium on Sea Buckthorn, October 19–23, 1989, Xi'an, China. p. 350–357.

Johansson, A., Lankso, P., and Kallio, H. 1997*a*. Characteristics of seed oils of wild, edible Finnish berries. Z. Lebensm. Unters Forsch. A. **204**: 300–307.

Johansson, A., Lankso, P., and Kallio, H. 1997*b*. Molecular weight distribution of the triacylglycerols of berry seed oils analysed by negative-ion chemical ionization mass spectrometry. Z. Lebensm. Unters Forsch. A. **204**: 308–315.

Johansson, A., Korte, H., Yang, B., Stanley, J., and Kallio, H.. 2000. Sea buckthorn berry oil inhibits platelet aggregation. J. Nutr.

Biochem. **11**: 491–495.

Kadamshoev, M. 1998. The green sea buckthorn aphid. Zash.-Karant. Rast. **12**: 22.

Kalinina, I.P. 1987. Breeding of sea buckthorn in the Altai. *In* Advances in agricultural science. *Edited by* A.B. Kryukov. Moscow, Russia. pp. 76–87. (from Plant Breeding Abst. **58**: 6090).

Kallio, H., Manninen, P., Haivala, E., Sarimo, S., and Korteniemi, V. 1995. Aseptic production of sea buckthorn oil capsules based on supercritical fluid extraction. Worldwide Research & Development of Sea buckthorn. *In* Proceedings of the International Workshop on Sea Buckthorn, December 1995, Beijing, China. p. 157–158.

Kallio, H., Yang, B., Tahvonen, R., and Hakala, M. 2000. Composition of sea buckthorn berries of various origins. *In* Proceedings of the International Workshop on Sea Buckthorn, December 1995, Beijing, China. p. 13–19.

Karhu, S.T., and Ulvinen, S.K. 1999. Vitamin C: a variable quality factor in sea buckthorn breeding. *In* Agri-Food Quality II: quality management of fruits and vegetables – from field to table. *Edited by* M. Hagg, R. Ahvenainen, and A.M. Evers. Second International Conference, Turku, Finland, 22-25 April 1998. pp. 360–367.

Kennedy, D.M. 1987. Verticillium wilt of sea buckthorn (*Hippophae rhamnoides*). Plant Pathol. **36**: 420–422.

Kesariiski, A.G., Nikolyuk, V.I., and Podgainyi, V.A. 1990. Sea buckthorn oil production by extracting mixtures of milled pressings and stones of buckthorn with vegetable oil, with simultaneous ultrasonic treatment. Soviet Union Patent 1750686.

Khaidarov, K., Rachimov, L., and Lebedeva, L.D. 1989. The influence of sea buckthorn oil on the organs of vital importance, motor- and secretory functions of the gastrointestinal tract. *In* Proceedings of the First International Symposium on Sea Buckthorn, October 19–23, 1989, Xi'an, China. pp. 399–400.

Kleinschmidt, T., Siudzinski, S., and Lange, E.

1996. Stabilization of the oil and cloud phases in sea buckthorn juice. Flussiges Obst. **63**: 702–705; Food Sci. Technol. Abst. 3H134, 1997.

Kluczynski, B. 1979. Suitability of selected tree and shrub species for the reclamation of ash wastes from power stations. Arbor. Kornickie **24**: 217–282.

Kluczynski, B. 1989. Effects of sea buckthorn (*Hippophae rhamnoides* L.) cultivation on post-industrial wastelands in Poland. *In* Proceedings of the First International Symposium on Sea Buckthorn, October 19–23, 1989, Xi'an, China. p.275–287.

Kniga, N.M. 1989. Characteristics of rooting softwood cuttings of top and small fruit species in relation to natural photoperiod. Fiziol. Biokhim. Kul't. Rast. 21: 403–409.

Koch, H.J. 1981. Cultivation of sea buckthorn for fruit production for the fruit processing industry. Gartenbau 28: 175–177.

Kondrashov, V.T. 1981*a*. Dieback of *Hippophae rhamnoides*. Lesn. Khoz. **7**: 50–53.

Kondrashov, V.T. 1981*b*. Structural elements of sea buckthorn productivity. Biol. Nauk. **7**: 81–85.

Kondrashov, V.T. 1986*a*. Study of yield in *Hippophae rhamnoides* in relation to breeding. Sostoyanie i perspektivy razvitiya kul'tury oblepikhi v Nechernozemnoi zone RSFSR. Materialy Soveshch., Moskva, 19 fevr., 1982. 23–27. (from Plant Breed. Abst. **57**: 11133).

Kondrashov, V.T. 1986*b*. The productivity of sea buckthorn varieties in relation to breeding. Sb.k Nauchn. Tr., Vses. Nauchno-Issled. Inst. Sadov. imeni I. V. Michurina **46**: 86–89. (from Plant Breed. Abst. **58**: 838).

Kostrikova, E.V. 1989. Experimental study of wound-healing effect of the preparation "Aekol" (artificial sea buckthorn oil). Ortop. Travmatol Prot. **1**: 32–36.

Kondrashov, V.T., and Kuimov, V.N. 1987. Vegetative propagation of *Hippophae rhamnoides*. Sadovodstvo **6**: 13–16. (from Hortic. Abst. **58**: 2047).

Kondrashov, V.T., and Sokolova, E.P. 1990. New wilt-resistant forms of *Hippophae rhamnoides.* Byull. Moskov. Obshch. Ispyt. Prir. Biol. **96**: 146–153. (from Plant Breed. Abst. **62**: 733).

Kondrashov, V.T. 1994. New technology of creating stands of sea buckthorn and other horticultural crops. Russian Agri. Sci. **11**: 24–28.

Kostyrko, D.R. 1990. Introduction of useful food plants into the Donetsk Botanic Garden of the Ukrainian Academy of Sciences. Introd. Akklimat. Rast. **14**: 31–34. (from Hortic. Abst. **61**: 3368).

Kudritskaya, S.E., Zagorodskaya, L.M, and Shishkina, E.E. 1989. Carotenoids of the sea buckthorn, variety Obil'naya. Chem. Nat. Compd. **25**: 724–725.

Kukenov, M.K., Dzhumagalieva, F.D., Tatimova, N.G., and Bespaev, S.B. 1982. Study of medicinal plant reserves and distribution atlas compilation in the Kazakh-SSR USSR and prospects of their use in public health service. Izvest. Akad. Nauk Kaz. SSR Ser. Biol. **1**: 3–6.

Kuznetsov, P.A. 1985. Effect of pre-planting treatment and plastic mulch on rooting of sea buckthorn hardwood cuttings and transplant quality. Biol. Aspek. Introd. Selek. Agrotekhn. Oblep. **1985**: 159–163. (from Hortic. Abst. **56**: 8687).

Lange, E., Klein, G., Gerber, J., Bauer, F., Fetkenhauser, W., and Sievert, B. 1991. A procedure for manufacturing semi-finished sea buckthorn products. Lebensmitteltechnik **25**: 37–40; as reported, Food Sci. Technol. Abst. **25**: 5T30.

Laurinen, E. 1994. Non-traditional culture of tree fruit and small fruit crops outside the normal season and new species for economic production. Nordi. Jordbruksforsk. **76**: 149–174.

Lebedeva, L., Rachmov, I., and Kchaidarov, K. 1989. Screening investigation of the antiinflammation activity of sea buckthorn oil. *In* Proceedings of the First International Symposium on Sea Buckthorn, October 19–23, 1989, Xi'an, China. pp. 398–399.

Li, J., Liu, X., and Chen, Z. 1996. Sea buckthorn and chinese stone nutrient liquor. Chinese Patent 1104054.

Li, R. 1990. A brief introduction to the techniques of raising sea buckthorn seedlings. Hippophae **1**: 18–19. [In Chinese]

Li, T.S.C., and Schroeder, W.R. 1996. Sea buckthorn (*Hippophae rhamnoides* L.): a multipurpose plant. HortTechnology **6**: 370–380.

Li, T.S.C., and Schroeder, W.R. 1999. A growers guide to sea buckthorn. Agric. Agri-Food Can. Publ. 70 p.

Li, T.S.C., and Wang, L.C.H. 1998. Physiological components and health effects of ginseng, *Echinacea*, and sea buckthorn. *In* Functional Foods. *Edited by* G. Mazza. Technomic Publ. Co. Inc. Lancaster, PA. pp. 329–356.

Li, T.S.C., and Wardle, D.A. 1999. Effects of seed treatmetns and planting depth on emergence of sea buckthorn species. HortTechnology **9**: 213–216.

Li, X. 1996. Technology for production of *Hippophae rhamnoides* seed oil. Chinese Patent 1123318.

Li, Y., and Xu, M. 1993. Preliminary report on the anti-bacterial effect of sea buckthorn oil. Hippophae **6**: 28–29

Li, Y. R., and Wang. L.Y. 1994. A preliminary analysis of the effects of sea buckthorn oil capsule and sea buckthorn 'Maisaitong' capsule on ischemic apoplexy. Hippophae **7**: 45–46. [In Chinese]

Lian, Y. 1988. New discoveries of the genus *Hippophae*. Acta Phytotaxon. Sin. **26**: 235–237.

Lian, Y., and Chen, S.L. 1997. Botany and Biology of Genus *Hippophae*. Int. Centre Res. Training Seabuckthorn Publication. Beijing, China. 105 p. [In Chinese]

Lian, Y., Lu, S.K., Zhen, S.K., and Chen, S.L. 2000. Biology and Chemistry of Plants of Genus *Hippophae*. GanSu Sci. Technology Publ. Co. GanSu, China. 228 p. [In Chinese]

Libman, M., and Zolotarsky, V. 2001. A composition including sea buckthorn oil extract and antioxidant and/or a UV filter. World Patent 0053152.

Liu, J., and Liu, Z. 1989. Research of processing technology for sea buckthorn concentrated juice. *In* Proceedings of the First International Symposium on Sea Buckthorn, October 19–23, 1989, Xi'an, China. pp. 314–317.

Liu, S.W., and He, T.N. 1978. The genus *Hippophae* from the Quin-Zang Plateau. Acta Phytotaxonomica **16**: 106–108. [In Chinese]

Lorber, S., and Heilscher, K. 1996. Production of sea buckthorn oil. German Patent 4431393.

Loskutova, G.A., Baikov, V.G., Starkov, A.V., and Medvedev, F.A. 1989. The composition of fatty acids from the lipids of *Hippophae rhamnoides* L. fruits. Rastit. Resur. **25**: 97–103. (from Hortic. Abst. **59**: 7303).

Lu, R. 1992. Sea buckthorn: A multipurpose plant species for fragile mountains. Int. Centre for Integrated Mountain Development, Katmandu, Nepal. 62 p.

Lu, S., and Ma, C. 2001. Direction , focus and contents of sea buckthorn research and development in China-facing the new century. *In* Proceedings of an International Workshop on Sea Buckthorn. A Resource for Health and Environment in Twenty First Century, February 18–21, 2001, New Delhi, India. pp. 6–11.

Lui, J., and Lui, Z. 1989. Research of processing technology for sea buckthorn concentrated juice. *In* Proceedings of the First International Symposium on Sea Buckthorn, October 19–23, 1989, Xi'an, China. pp. 318–319.

Ma, M., Cui, C., and Feng, G. 1989. Studies on the fruit character and biochemical compositions of some forms within Chinese sea buckthorn (*Hippophae rhamnoides* subspecies sinensis) in Shanxi, China. *In* Proceedings of the First International Symposium on Sea Buckthorn, October 19–23, 1989, Xi'an, China. pp. 106–113.

Ma, Z., and Sun, H. 1986. Interrelationship between sea buckthorn and some birds and beasts. J. Ecology **5**: 30–32. [In Chinese]

Mackay, J., Simon, L., and Laionde, M. 1987. Effect of substrate on the performance of in vitro propagated *Alnus glutinosa* clones inoculated with Sp+ and Sp- *Frankia* strains. Plant Soil **103**: 21–31.

Magherini, R. 1986. Considerations on the biological potential of *Hippophae rhamnoides* L. *In* Atti. Convegno sulla Coltivazione delle Piante Officinali, Trento, October 9–10, 1986. pp. 397–410. (from Hortic. Abst. **58**: 6533).

Makinen, K.K., and Soderling, E. 1980. A quantitative study of mannitol, sorbitol, xylitol and xylose in wild berries and commercial fruits. J. Food Sci. **45**: 367–374.

Mamedov, S.Sh., Gigienova, E.I., Umarov, A.U., and Aslanov, S.M. 1981. Oil lipids in the fruits and leaves of *Hippophae rhamnoides*. Khim. Prir. Soedin. **6**: 710–715.

Mann, D.D, Petkau, D.S., Crowe, T.G., and Schroeder, W.R. 2000. Removal of sea buckthorn (*Hippophae rhamnoides* L.) berries by shaking. Ann. Meeting Canadian Hortic. Soc., July 15–19, 2000, Winnipeg, MB.

Mann, D.D, Petkau, D.S., Crowe, T.G., and Schroeder, W.R. 2001. Removal of sea buckthorn (*Hippophae rhamnoides* L.) berries by shaking. Can. Biosyst. Eng. **43**: 223–228.

Manninen, P., Leakso, P., and Kallio, H. 1995. Method for characterization of triacylglycerols and fat soluble vitamins in edible oils and fats by supercritical fluid chromatography. J. Am. Oil Chem. **72**: 1001–1008.

Martem'yanov, P.B., and Khromova, T.V. 1985. Agrotechnical measures for advancing woody plant growth. Byull. Glavn. Bot. Sada **138**: 45–48. (from Hortic. Abst. **56**: 8066).

Meireles, M.A.A., and Nikolov, Z.L. 1994. Extraction and fractionation of essential oils with liquid carbon dioxide. *In* Spices, Herbs and Edible Fungi. *Edited by* G. Charlambous. Elsevier Science B.V., Amsterdam, London, New York. p 171–199.

Messerschmidt, K., Raasch, A., and Knorr, D. 1993. Colors from waste products. Extraction of natural plant pigments from sea buckthorn using supercritical CO_2. Food Sci. Technol. Abstr. **25**(5): 5T30.

Mironov, V.A. 1989. Chemical composition of *Hippophae rhamnoides* of different populations of the USSR. *In* Proceedings of the First International Symposium on Sea Buckthorn, October 19–23, 1989, Xi'an, China. pp. 67–70.

Mironov, V.A., Vasilev, G.S., Matrosov, V.S., Muzychenko, L.D., Usha, B.V., Kasyanenko, I.I., and Feldshtein. M.A. 1980. Seabuckthorn oil obtained by the extraction method and its biological activity. Khim-Farm. Zh. **14**: 74–80. [In Russian]

Mironov, V. A., Guseva-Donskaya, T.N., Dubrovina, Yu Yu, Osipov, G.A., Shabanova, E.A., Nikulin, A.A., Amirov, N.Sh., and Trubitsina, I.G. 1989. Chemical composition and biological activity of extracts from sea buckthorn fruit compoents. Khim-Farm. Zh. **23**: 1357–1364. [In Russian]

Mironov, V.A., Guseva-Donskaya, T.N., Dubrovina, Y., Osipova, G.A., Shabanova, E.A., Nikulin, A.A., Amirov, N.S., and Trubitsina, I.G. 1991. Composition and biological activity of lipid extracts from Armenian sea buckthorn. In Nov. Biol. Khim. Farm. Oblep. AN SSSR, Novosibirsk, Russia. pp. 114–121. [In Russian]

Mishulina, I.A. 1976. The effect of foliar nutrition with minor elements of *Hippophae rhamnoides* on the fruit content of biologically active substances. Biol. Aktiv. Vesh. Plodov i Yagod (1976) 97–99. (from Hortic. Abst. **48**: 320).

Montpetit, D., and Lalonde, M. 1988. In vitro propagation and subsequent nodulation of the actinorhizal Hippophae rhamnoides L. Plant Cell Tissue Organ Cult. (Netherlands) **15**: 189–200.

Morar, R., Cimpeanu, S., Morar, E., Marghitas, L., and Rozalia, Z. 1990. Results of the use of certain phytotherapeutic preparations in the feeding of weaned piglets. Bul. Inst. Agron. Cluj-Napoca Ser. Zooteh. Med. Vet. **44**: 101–108.

Müller, K.D. 1993. Variability of selected characters of buckthorn clones and its importance. *In* Cultivation and utilization of wild fruit crops. Bernhard Thalacker Verlag GmbH & Co. p. 51–56. [In German]

Müller, P. 1995. Verbreitungsbiologie der Blutenpflanzen. Berlin, Germany. 152 p.

Myakushko, V.E., Kosenko, V.M., and Bedritskii, A.S. 1986. *Hippophae rhamnoides* in stands of gulley and ravine systems. Lesn. Khoz. **10**: 30–34.

Nizhegorodtsev, Y. M., and Umanskii, M.S. 1991. Production of concentrate of sea buckthorn oil involves freezing and thawing sea buckthorn berries, separating juice and flesh, drying and centrifuging. Russian Patent 2076899.

Novruzov, E.N., and Aslanov, S.M. 1983. Studies on the dynamics of ascorbic acid accumulation in sea buckthorn fruits. Dokl. Akad. Nauk Az. SSR **39**: 59–63. (from Hortic. Abst. **54**: 8057).

Olander, S. 1995. Mechanical harvesting of sea buckthorn. *In* Proceedings of the International Workshop on Sea Buckthorn, December 1995, Beijing, China.

Oomah, B.D., Sery, G., Godfrey, D.V, and Beveridge, T.H.J. 1999. Rheology of sea buckthorn (*Hippophae rhamnoides* L.) juice. J. Agric. Food Chem. **47**: 3546–3550.

Osipov, Y.V. 1983. Propagation of sea buckthorn by cuttings. Sadovodstvo **12**: 20–21.

Ozerinina, O.V., Berezhnaya, G.A., and Vereshchagin, A.G. 1987.Triacylglycerol composition and structure of sea buckthorn fruits grown in different regions. Russian J. Plant Physiol. **44**: 62–69.

Pan, R.Z., Zhang, Z., Ma, Y., Sun, Z., and Deng, B. 1989. The distribution characters of sea buckthorn (*H. rhamnoides* L.) and its research progress in China. In Proceedings of the First International Symposium on Sea Buckthorn, October 19–23, 1989, Xi'an, China. p. 1–16.

Papanikolaw, J. 1999. Sea buckthorn oil poised to grow in the US nutraceutical market. Chem. Market Rep., Sept. 6, 1999.

Pearson, M.C., and Rogers, J.A. 1962. Biological flora of the British Isles. No.85. *Hippophae rhamnoides* L. J. Ecol. **50**: 501–509.

Perez-Jimenez, F., Castro, P., Lopez-Miranda, J., Paz-Rojas, J., Blanco, E., Lopez-Segura,

A., Velasco, F., Marin, C., Fuentes, F., and Ordovas, J.M. 1999. Circulating levels of endothelial function are modulated by dietary mono-unsaturated fats. Atherosclerosis **145**: 351–358.

Petrova, O.P. 1982. Fungal wilt pathogen of *Hippophae rhamnoides*. Bull. Glavnogo Botanicheskogo Sada **124**: 96–97.

Piir, R. 1996. *Hippophae rhamnoides* L. in Estonia for fruit growing. Agraarteadus **7**: 162–176.

Pintea, A., Marpeau, A., Faye, M., Socaciu, C., and Gleizes, M. 2001. Polar lipid and fatty acid distribution in carotenoipoprotein complexes extracted from sea buckthorn fruits. Phytochem. Anal. 2001. **12**: 293–298.

Polikarpova, F.Y., Upadyshev, M.T., and Oskareva, G.P. 1999. Propagation of berry shrubs and some fruit shrubs by semi-lignified and lignified leafy cuttings. Sadovod. Vinograd. **2**: 18–20.

Predeina, R.V. 1987. Fertilization of *Hippophae rhamnoides* in the Altai region. Sadovodstvo 6: 13–18. (from Hortic. Abst. **58**: 2046).

Prokof, E.P.S., Sergeev, A.V., Uteshev, B.S., Kostrjukov, E.B., Kovaleva, V.L, and Storozhakov, G.I. 1999. Ointment for treatment of patient with inflammatory and allergic skin damages. Russian Patent 2132183.

Quirin, K.W., and Gerard, D. 1993. Sanddornlipide -interessante Wirkstoffe fü r die Kosmetik. Parfümerie Kosmetik **10**: 618–625.

Rachimov, I.P., Lebegdeva, L.D., and Khaidarov, K.K. 1989. The experimental toxicology of the sea buckthorn oil. *In* Proceedings of the First International Symposium on Sea Buckthorn, October 19–23, 1989, Xi'an, China. pp. 371–372.

Ranwell, D.S. 1972. The ecology of *Hippophae* within the dune system. *In* The Management of Sea Buckthorn (*Hippophae rhamnoides* L.) on Selected Sites in Great Britain. *Edited by* D.S. Ranwell. Nature Conservancy Council. Oxford Publ. Blackwell, Great Britain. pp. 22–27.

Ranwell, D.S. 1979. Strategies for the management of coastal systems in ecological processess. *In* Coastal Environments. *Edited by* R.L. Jeffries and A.J. Davy. Nature Conservancy Council. Oxford Publ. Blackwell, Great Britain. p. 515–527.

Ridley, H.N. 1930. The Dispersal of Plants Throughout the World. Ashford, Kent, U.K. 744 p.

Rizvi, S.S.H., Benado, A.L., Zollweg, J.A., and Daniels, J.A. 1986. Supercritical fluid extraction: fundamental principles and modeling methods. Food Technol. **42**(6): 55–65.

Robinson, N.A. 1972. The control of Hippophae. *In* The Management of Sea Buckthorn (*Hippophae rhamnoides* L.) on selected sites in Great Britain. *Edited by* D.S. Ranwell. Nature Conservancy Council. Oxford Publ. Blackwell, Great Britain. pp. 28–35.

Rousi, A. 1965. Observations on the cytology and variation of European and Asiatic populations of *Hippophane rhamnoides*. Ann. Bot. Fenn. **2**: 1–18.

Rousi, A. 1971. The genus *Hippophae* L. A taxonomic study. Ann. Bot. Fenn. **8**: 177–227.

Rousi, A., and Aulin, H. 1977. Ascorbic acid content in relation to ripenesss in fruit of six *Hippophae rhamnoides* clones from Pyharanta, SW Finland. Ann. Agric. Fenn. **16**: 80–87.

Rui, L., Gao, Y., and Su, R. 1989. Effects of seabuckthorn oil on lipids peroxidation of guinea pigs erythrocyte membranes. *In* Proceedings of the First International Symposium on Sea Buckthorn, October 19–23, 1989, Xi'an, China. pp. 358–364.

Saint-Cricq de Gauljac, N., Provost, C., and Vivas, N. 1999. Comparative study of polyphenol scavenging activities assessed by different methods. J. Agric. Food Chem. **47**: 425–431.

Salenko, V.L., Kukina, T.P., Karamyshev, V.N., Sidelnikov, V.N., and Pentagova, V.A. 1985. Chemical investigation of *Hippophaë rhamnoides*. II. Main components of the neutral reaction of the saponification products of an extract of the leaves of sea buckthorn. Translated from Khim. Prir. Soedin. **4**: 514–519.

Salo, K. 1991. The initial development of the

Chinese sea buckthorn, *Hippophae rhamnoides* subsp. *Sinensis* in the greenhouse. Luonnon Tutkija **95**: 150–155. [In Finnish]

Savkin, V.A., and Mukhamadiev, S.M. 1983. Effect of branch detachment from sea buckthorn bushes on their regeneration and fruiting. Ekol. Rast. Zhivot. Zapovedn. Uzbek. (1983) 29–32. (from Hortic. Abst. **54**: 8055).

Schapiro, D.C. 1989. Biochemical studies on some hopeful forms and species of sea buckthorn in USSR. *In* Proceedings of the First International Symposium on Sea Buckthorn, October 19–23, 1989, Xi'an, China. p.64–66.

Schroeder, W.R. 1988. Planting and establishment of shelterbelts in humid severe-winter regions. Agric. Ecosyst. Environ. **22/23**: 441–463.

Schroeder, W.R. 1990. Shelterbelt planting in the Canadian prairies. In: Protective plantation technology. Publishing House of Northeast Forestry Univ., Harbin, China. p.#35–43.

Schroeder, W.R. 1995. Improvement of conservation trees and shrubs. PFRA Shelterbelt Centre, Indian Head, SK, Canada. Suppl. Rep. #95-1, 42 p.

Schroeder, W.R., and Yao, Y. 1995. Sea buckthorn: a promising multipurpose crop for Saskatchewan. Prairie Farm Rehabilitation Administration, Agriculture & Agri-Food Canada. 10 p.

Senjavina, N.K., Sorokin, N.I., Tarasov, A.V., Babaev, V.V., Chernykh, I.N., Davydova, V.N., Gordievskij, A.A., Kozhevinkov, B.E., Li, S.K., Petukhov, V.A., and Sekretev, A.G. 1998. Skin care means. Russian Patent 2106859.

Shaftan, E.A., Mikhailova, N.S., Gol'berg, N.D., and Pekhov, A.V. 1986. An investigation of the chemical composition of a CO_2 carbon dioxide extract from the pulp of *Hippophae rhamnoides*. Chem. Nat. Compd. **14**: 560–561.

Shchapov, N.S., and Kreimer, V.K. 1988. Experimental polyploids of sea buckthorn (*Hippophae rhamnoides* L.) I. Producing and identifying polyploids. Invest. Sibirsk. Otdel. Akad. Nauk SSR Biol. Ikh Nauk **6**: 111–117.

Siabaugh, P.E. 1974. *Hippophae rhamnoides*–Common sea buckthorn. *In* Seeds of woody plants in the United States. *Technical Co-Ordinator*: C.S. Schopmeyer. USDA-FS Agric. Hdbk No. **450**: 446–447.

Siimisker, T. 1996. Sea buckthorn cultivar evaluation at the southern Estonian cultivar testing farm. Agraarteadus. **7**: 177–181.

Singh, V. 2001. Sea buckthorn (*Hippophae* L.)–A wonder plant of dry temperate Himalayas. *In* Proceedings of an International Workshop on Sea Buckthorn. A Resource for Health and Environment in Twenty First Century, February 18–21, 2001, New Delhi, India. p.39–42.

Skogen, A. 1972. The Hippophae rhamnoides Alluvial forest at Leinora, central Norway. A phytosociaological and ecological study. K. Norske Vidensk. Selsk. Skr. **4**: 1–115.

Small, E., Catling, P.M., and Li, T.S.C. 2002. Blossoming treasures of biodiversity: 5. Sea buckthorn (*Hippophae rhamnoides*)–an ancient crop with modern virtues. Biodiversity 3: 25–27.

Smirnova, N.G., and Tikhomirova, N.I. 1980. Combined use of X-ray photography and the tetrazolium method for assessing seed viability. Byull. Glavn. Bot. Sada **117**: 81–85. (from Hortic. Abst. **51**: 6430).

Sokoloff, B.K., Funaoka, M., Fujisawa, C.C., Saelhof, E., Raniguchi, D.B., and Miller, C. 1961. An oncostatic factor present in the bark of *Hippophae rhamnoides*. Growth **25**: 401–409.

Solonenko, L.P., and Shishkina, E.E. 1983. Proteins and amino acids in sea buckthorn fruits. Biologiya, Khimiya i Farmakologiya Oblepikhi 1983, 67–82.

Stastova, J., Jfz, J., Bartlova, M, and Sovova, H.. 1996. Rate of the vegetable oil extraction with supercritical CO_2–III. Extraction from sea buckthorn. Chem. Eng. Sci. **51**: 4347–4352.

Stewart, W.D.P., and Pearson, M.C. 1967.

Nodulation and nitrogen-fixation by *Hippophae rhamnoides* L. in the field. Plant Soil **26**: 348–360.

Sun, Z. 1995. Exploitation and utilization of sea buckthorn (*H. rhamnoides* L.) in China. NorthWest University Publication, ShiAn, China. 99 p.

Synge, P.M. 1974. Dictionary of gardening: a practical and scientific encyclopaedia of horticulture. 2nd ed. Clarendon Press, Oxford. 2316 p.

Tang, X. 2002. Intrinsic change of physical and chemical properties of sea buckthorn (*Hippophae rhamnoides*) and implications for berry maturity and quality. J. Hortic. Sci. Biotech. **77**: 177–185.

Temelli, F., Chen, C.S., and Braddock, R.J. 1988. Supercritical fluid extraction in citrus oil processing. Food Technol. **42**(6):145–150.

Thompson, J.R., and Rutter, A.J. 1986. Effects of sodium chloride on some native British shrub species, and the possibility of establishing shrubs on the central reserves of motorways. J. Appl. Ecol. **23**: 299–315.

Thurnham, D.I. 1999. Functional foods: cholesterol-lowering benefits of plant sterols. Br. J. Nutr. **82**: 255–256.

Tong, J., Zhang, C., Zhao, Z., Yang, Y., and Tian, K. 1989. The determination of the physico-chemical constants and sixteen mineral elements in sea buckthorn raw juice. *In* Proceedings of the First International Symposium on Sea Buckthorn, October 19–23, 1989, Xi'an, China. p. 132–137.

Triebold, H.O., and Aurand, L.W. 1963. Food Composition and Analysis. D. Van Norstrand Co. Ltd., New York, London, Toronto. 253 p.

Trunov, I.A. 1996. Removal of nutrients by plants in greenhouse. Sadovod. Vinograd. **2**: 12.

Trushechkin, V.G., Lobanova, G.V., Ostreiko, S.A., Kalinina, I.P., Bartenev, V.D., and Burdasov, V.M. 1973. The use of beta-chloroethylphosphonic acid for facilitating the detachment of *Hippophae rhamnoides* and *Aronia melanocarpa* fruit in relation to mechanized harvesting. Sb. Nauchn. Rabot, Nauchn.-Issled. Zon. Inst. Sadov. Nechern. Pol. **6**: 168–173. (from Hortic. Abst. **45**: 3902).

Turpein, A.M., Pajari, A.M., Freese, R., Sauer, R., and Mutanen, M. 1998. Replacement of dietary saturated fatty acids by unsaturated fatty acids: effects on platelet protein kinase C activity, urinary content of 2,3-dinor-TXB2 and in vitro platelet aggregation in healthy man. Thromb. Haemostasis **80**: 649–655.

Varlamov, G.P., and Gabuniya, V.G. 1990. Picking sea buckthorn fruit by suction air stream. Trak. Sel'sk. Mash. **1**: 29–30.

Velioglu, Y.S., Mazza, G., Gao, L., and Oomah, B.D. 1998. Antioxidant activity and total phenolics in selected fruits, and vegetables, and grain products. J. Agric. Food Chem. **46**: 4113–4117.

Verehschagin, A.G., Ozerinina, O.V., and Tsydendambaev, V.D. 1998. Occurrence of two different systems of triacyglycerol formation in sea buckthorn fruit mesocarp. J. Plant Physiol. **153**: 208–213.

Vereshchagin, A.G., and Tsydendambaev, V.D. 1995. Neutral lipids of mature and developing sea buckthorn (*Hippophae rhamnoides* L) fruits. *In* Plant Lipid Metabolism. *Edited by* J.C. Kader and P. Mazliak. Kluwer Academic Publ., Dordrecht, NL. pp. 573–579.

Vernik, R.S., and Zhapakova, U.N. 1986. Biology of seed germination in *Hippophae rhamnoides*. Uzbek. Biol. Zh. 5: 38–40.

Vlasov, V.V. 1970. Hippophae oil in the treatment of superficial skin burns. Vestn. Dermatol. Veneraol. **44**: 69–72.

Wahlberg, K., and Jeppsson, N. 1990. Development of cultivars and growing techniques for sea buckthorn, black chokeberry, *Lonicera*, and *Sorbus*. Sver. Lantbruksuniv. Verksamhetsb. Balsgaard (Sweden) 1990; 1988–1989: 80–93.

Wang, B., Feng, Y., Yu, Y., Zhang, H., and Zhu, R. 2001. Effects of total flavones of *Hippophae rhamnoides* L. (sea buckthorn) on cardiac function and hemodynamics in healthy human subjects. Translation from the original Chinese provided by Rich Nature Nutroceutical Laboratories, Inc. [http://www.richnature.com/products/herbal/articles/heart.pdf]

Wang, H., Ge, H., and Zhi, J. 1989. The components of unsaponifiable matters in sea buckthorn fruit and seed oil. *In* Proceedings of the First International Symposium on Sea Buckthorn, October 19–23, 1989, Xi'an, China. pp. 81–90.

Wang, S. 1990. Studies on the chemical components in fruits of *Hippophae rhamnoides*. Forest Res. **3**: 98–102.

Wang, S. Z. 1990. Research on the chemical composition of three subspecies of *Hippophae rhamnoides* L. Forest Res. **3**: 98–102. [In Chinese]

Wei, X., and Cao, Z. 1995. Medicine for regulating endocrine function of woman. Chinese Patent 1106691.

Weiss T.J. 1963. Fats and oils. *In* Food Processing Operations, Their Management, Machinery, Materials, and Methods. Vol. 2. *Edited by* J.L. Herd and M.A. Joslyn. AVI Publ. Co. Inc., Westport, CT. p. 87–138.

Weiss, T.J. 1970. Food Oils and Their Uses. Chapter 3. Basic Processing of Fats and Oils. AVI Publ. Co. Inc., Westport, CT. pp. 47–67.

Wolf, D., and Wegert, F. 1993. Experience gained in the harvesting and utilization of sea buckthorn. *In* Cultivation and utilization of wild fruit crops. Bernhard Thalacker Verlag GmbH & Co. pp. 23–29. [In German]

Wu, A., Su, Y., Li, J., Liu, Q., Lu, J., Wei, X., Cian, C., Lai, Y., and Wang, G. 1989. The treatment of chronic cervicitis with *Hippophae* oil and its suppository (129 cases analysis). *In* Proceedings of the First International Symposium on Sea Buckthorn, October 19–23, 1989, Xi'an, China. pp. 404–406.

Xin, Y., and Zhou, B. 1994. Processing method for *Hippophae rhamnoides* seed oil. Chinese Patent 1089301.

Xin, Y., Li, Y., Wu, S., and Sun, X. 1995. Aitzmuller, K. A study of the composition of sea buckthorn oils in China. *In* Proceedings of the International Workshop on Sea Buckthorn, December 1995, Beijing, China. pp. 82–89.

Xing, G., and Wang, S. 2000. Sea buckthorn essence concentrate. Chinese Patent 1260170.

Xu, M. 1994. The medical research and exploitation of sea buckthorn. Hippophae 7: 32–84. [In Chinese]

Xu, M., and Dai, Y.C. 1997. A new forest pathogen of *Hippophae* in China: *Phellinus hippophaeicola*. Forest Res. **10**: 380–382.

Xu, M., Qian, Z.H., and Sun, P. 1989. A survey of medical research of *Hippophae rhamnoides* L. in China. *In* Proceedings of the First International Symposium on Sea Buckthorn, October 19–23, 1989, Xi'an, China. pp. 329–332.

Xu, M., Sun, X., and Cui, J. 2001. The medicinal research on sea buckthorn. *In* Proceedings of an International Workshop on Sea Buckthorn. A Resource for Health and Environment in Twenty First Century, February 18–21, 2001, New Delhi, India. p. 12–23

Xu, Q., and Chen, C. 1991. Effects of oil of *Hippophae rhamnoides* on the experimental thrombus formation and blood coagulation system. Res. Dev. Nat. Prod. **3**(3): 70–73. [In Chinese]

Yamori, Y., Nara, Y., Tsubouchi, T., Sogawa, Y., Ikeda, K., and Horie, R. 1986. Dietary prevention of stroke and its mechanism in stroke-prone spontaneously hypertensive rats, Preventive effect of dietary fiber and palmitoleic acid. J. Hypertens. **4** (Suppl. 3): 449–452.

Yang, B., Kalimo, K.O., Tahvonen, R.L., Mattila, L.M., Katajisto, J.K., and Kallio, H.P. 2000. Effects of dietary supplementation with sea buckthorn (*Hippophae rhamnoides*) seed and pulp oils on fatty acid composition of skin glycerophospholipids of patients with atopic dermatitis. J. Nutr. Biochem. **11**: 338–340.

Yang, B., and Kallio, H.P. 2001. Fatty acid composition of lipids in sea buckthorn (*Hippophaë rhamnoides* L.) berries of different origins. J. Agric. Food Chem. **49**:

1939–1947.

Yang, B., Karlsson, R.M., Oksman, P.H., and Kallio, H.P. 2001. Phytosterols in sea buckthorn (*Hippophae rhamnoides* L.) berries: identification and effects of different origins and harvesting times. J. Agric. Food Chem. **49**: 5620–5629.

Yang, B., Kalimo, K.O., Mattila, L.M., Kallio, S.E., Katajisto, J.K., Peltola, O.J., and Kallio, H.P. 1999. Effects of dietary supplementation with sea buckthorn (*Hippophae rhamnoides*) seed and pulp oils on atopic dermatitis. J. Nutr. Biochem. **10**: 622–630.

Yang, H. 1989. The dynamic distribution of some active components in sea buckthorn fruits growing on Qinghai-Tibet plateau. *In* Proceedings of the First International Symposium on Sea Buckthorn, October 19–23, 1989, Xi'an, China. p. 158–162.

Yang, H.Z., Liu, Y.L., Huo, S.H., and Zhang, G.L. 1992. The dynamic changes in the oil content and fatty acid composition of sea buckthorn fruit during the ripening period. J. Beijing Forestry Univ. **14**(2): 68–73. [In Chinese]

Yang, J., Wang, X., Liu, Y., Li, G., Ren, L., Jing, J., Zhang, H., Zhong, C., and Su, R. 1989. Preliminary studies on the effects of oil from fruit residues of sea buckthorn upon anti-tumors. In Proceedings of the First International Symposium on Sea Buckthorn, October 19–23, 1989, Xi'an, China. pp. 382–384.

Yao, Y. 1994. Genetic diversity, evoluation and domestication in sea buckthorn (*Hippophae rhamnoides* L.) Ph.D. dissertation, University of Helsinki, Helsinki, Finland.

Yao, Y., and Tigerstedt, P.M.A. 1994. Genetic diversity in *Hippophae* L. and its use in plant breeding. Euphytica **77**: 165–169.

Yao, Y., and Tigerstedt, P.M.A. 1995. Geographical variation of growth rhythm, height, and hardiness and their relations in *Hippophae rhamnoides*. J. Am. Soc. Hortic. Sci. **120**: 691–698.

Yao, Y., Tigerstedt, P.M.A, and Joy, P. 1992. Variation of vitamin C concentration and character correlation between and within natural sea buckthorn (*Hippophae rhamnoides* L.) populations. Acta Agric. Scand. **42**: 12–17.

Yaonian, X., Yonghai, L., Sulin, W., Xiuzhi, S., and Aitztmuller, K. 1995. A study of the compositions of seabuckthorn oils in China. *In* Proceedings of the International Workshop on Sea Buckthorn, December 1995, Beijing, China. pp. 82–89.

Zadernowski, R., Nowak-Polakowska, H., Lossow, B., and Nesterowicz, J. 1997. Sea buckthorn lipids. J. Food Lipids **4**: 165–172.

Zham'Yansan, Ya. 1978. Investigation of seed and mesocarp oils of *Hippophaë rhamnoides* fruits. Khim. Prir. Soedin. **1**: 133–134.

Zhang, F., Gao, J., and Guo, Y. 1989. Predication of medical application prospects of sea buckthorn oil on the basis of its recent advances. *In* Proceedings of the First International Symposium on Sea Buckthorn, October 19–23, 1989, Xi'an, China. pp. 339–347.

Zhang, M. 1987. Random control tests of treating the ischemic cardiopathy with TFH. J. China Cardiovasc. Dis. **15**: 97–99.

Zhang, P. 1989. Anti-cancer activities of sea buckthorn seed oil and its effects on the weight of immune organs. Seabuckthorn **2**(3): 31–34.

Zhang, P., Ding, X., Mao, L., Li, D., and Li, L. 1989. Anti-tumor effects of fruit juice and seed oil of *Hippophae rhamnoides* and their influences on immune function. *In* Proceedings of the First International Symposium on Sea Buckthorn, October 19–23, 1989, Xi'an, China. p.373–381.

Zhang, W. 1988. Preliminary results of the experimental observation and the clinical application of treating the acute radio-dermatitis with sea buckthorn. Seabuckthorn **1**(1): 27–30

Zhang, W., Li, C., Liu, D., Deng, X., and Zhang, H. 1989*a*. Studies on ripening pattern of sea buckthorn (Hippophae) fruit. Acta Hortic. Sin. **16**: 241–247.

Zhang, W., Yan, J., Duo, J., Ren, B., and Guo, J. 1989*b*. Preliminary study of biochemical constituents of berry of sea buckthorn growing in Shanxi province and their changing trend. *In* Proceedings of the First International Symposium on Sea Buckthorn, October 19–23, 1989, Xi'an, China. pp. 96–105.

Zhao, J., Zhang, C.Y., Xu, Y., Huang, G.Q., Xu, Y.L., Wang, Z.Y., Fang, S.D., Chen, Y., and Gu, Y.L. 1990. The antiatherogenic effects of components isolated from pollen typhae. Thromb. Res. **38**: 957–966.

Zhao, Y. 1994. Preliminary report on the effects of sea buckthorn oil on thirty-two cases of burn or scald. Hippophae **7**(3): 36–37. [In Chinese]

Zhao, Z., Tong, J., Tang, T., Feng, G., Fan, M., and Shang, W. 1989. The determination of selenium in sea buckthorn raw juice with a method of automated fluorimetirc extraction. *In* Proceedings of the First International Symposium on Sea Buckthorn, October 19–23, 1989, Xi'an, China. p. 129–132.

Zhen, H. J., Chen, X.Y., Yang, Q.Z., and He, F.C. 1996. Effects of sea buckthorn oil on immune function of mice. J. Lanzhou Univ. (Nat. Sci.) **26**: 95–98. [In Chinese]

Zhmyrko, T.G., Goncharova, N.P., Gigienova, E.I., and Glushenkova, A.I. 1984. Group composition of neutral lipids in the oil from *Hippophae rhamnoides* fruit. Khim. Prir. Soedin. **3**: 300–305.

Zhmyrko, T.G., Glushenkova, A.I., Zegel'man, A.B., and Andronov, V.A. 1987. Neutral lipids in Freon extracts of *Hippophae rhamnoides* fruits. Khim. Prir. Soedin. **2**: 231–234.

Zhong, F. 1989. Study on the immunopharmacology of the components extracted from *Hippophae rhamnoides* L. *In* Proceedings of the First International Symposium on Sea Buckthorn, October 19–23, 1989, Xi'an, China. pp. 368–370.

Zhou, X., and Chen, C. 1989. Research on the process technology of turbidity type sea buckthorn beverage. *In* Proceedings of the First International Symposium on Sea Buckthorn, October 19–23, 1989, Xi'an, China. pp. 310–313.

Zhou, Y., and Jiang, J. 1989. Medical and health-care functions and applications of sea buckthorn. Hippophae **2**(2): 35–42. [In Chinese]

Zhou, Y., Ruan, D., Yang, B., and Wang, S. 1991. A study on vitamin C content of *Hippophae rhamnoides* fruit and its changing roles. For. Res. **4**: 345–349.

Zou, G. 1997. Multifunctional medicinal super cleansers. Chinese Patent 1144266.

Zou, X. 1999. Composite sea buckthorn oil capsule. Chinese Patent 1207920.

Index